The Conservation Reserve Program

Strategies, Impacts, and Environmental Benefits

snova
New York

Agriculture Issues and Policies Series

Introduction to Rural Sociology and Educational Psychology
Sasmitha R, PhD,
A. Muhammed Iqshanullah, PhD,
T. Kungumaselvan, PhD,
M. R. Naveen Kumar, PhD,
and Rajasekaran R, PhD
ISBN: 979-8-89530-734-2

Irrigation Management: Strategies, Sustainability and Agricultural Advancements
Maha Elbana, PhD,
and Youssef Brouziyne, PhD
ISBN: 979-8-89530-318-4
eBook ISBN: 979-8-89530-532-4

Tissue Culture Techniques in Vegetable Crop Improvement
Puja Rattan, PhD,
and Sanjeev Kumar, PhD
ISBN: 979-8-89530-049-7
eBook ISBN: 979-8-89530-073-2

Nutrition, Quality and Biological Potential of Grapevine
Marina T. Stojanova, PhD,
Dragutin Djukic, PhD,
Monika Stojanova, PhD,
and Aziz Şatana, PhD
ISBN: 979-8-89113-843-8
eBook ISBN: 979-8-89113-918-3

Agricultural Policy: Strategies, Challenges and Global Implications
Youssef Lee (Editor)
ISBN: 979-8-89113-720-2
eBook ISBN: 979-8-89113-787-5

The Future of Smart Agriculture
Hitesh Mohapatra, PhD,
Bhabendu Kumar Mohanta, PhD,
Soumya Ranjan Mishra, PhD
eBook ISBN: 979-8-89113-365-5

More information about this series can be found at
https://novapublishers.com/product-category/series/agriculture-issues-and-policies/

Jerry P. Norton
Editor

The Conservation Reserve Program

Strategies, Impacts, and Environmental Benefits

snova
New York

Library of Congress Cataloging-in-Publication Data

ISBN: 979-8-89530-774-8 (Hardcover)
ISBN: 979-8-89530-946-9 (eBook)

Published by Nova Science Publishers, Inc. † New York

Contents

Preface

The Conservation Reserve Program (CRP) is a land conservation initiative administered by the Farm Service Agency (FSA), designed to balance agricultural productivity with environmental stewardship. By enrolling in the CRP, farmers play a pivotal role in combating climate change, enhancing biodiversity, and preserving ecosystems for future generations. In exchange for a yearly rental payment, farmers enrolled in the program agree to remove environmentally sensitive land from agricultural production and plant species that will improve environmental health and quality. Contracts for land enrolled in CRP are from 10 to 15 years in length. The long-term goal of the program is to re-establish valuable land cover to help improve water quality, prevent soil erosion, and reduce the loss of wildlife habitat.

Chapter 1

Conservation Reserve Program*

U.S. Department of Agriculture

About the Conservation Reserve Program (CRP)

CRP is a land conservation program administered by the Farm Service Agency (FSA). In exchange for a yearly rental payment, farmers enrolled in the program agree to remove environmentally sensitive land from agricultural production and plant species that will improve environmental health and quality. Contracts for land enrolled in CRP are from 10 to 15 years in length. The long-term goal of the program is to re-establish valuable land cover to help improve water quality, prevent soil erosion, and reduce loss of wildlife habitat.

Created in 1985, CRP is one of the largest private-lands conservation programs in the United States. Thanks to voluntary participation by farmers and landowners, CRP has achieved significant goals in conservation and sustainability.

CRP enrollment options include General CRP, Grassland CRP and Continuous CRP. The Continuous CRP signup also includes additional options, such as CLEAR30, State Acres For Wildlife Enhancement (SAFE), Conservation Reserve Enhancement Program, Farmable Wetlands Program and Forest Management Incentive (FMI).

* This is an edited, reformatted and augmented version of U.S. Department of Agriculture Publication.

In: The Conservation Reserve Program
Editor: Jerry P. Norton
ISBN: 979-8-89530-774-8

News

FSA is accepting offers for more than 2.2 million acres from agricultural producers and private landowners through this year's Grassland, General, and Continuous CRP signups. With these accepted acres, enrollment is very near the 27 million CRP acreage cap. Producers can still make an offer to participate in CRP through the Continuous CRP signup, which is ongoing, by contacting FSA at their local USDA Service Center. Read more in our August 19 news release.

Sign up for Grassland CRP ran June 3 - 28, 2024. Grassland CRP is a working lands program – helping producers and landowners protect grassland from conversion while enabling haying and grazing activities to continue. Lands enrolled support haying and grazing operations and promotes plant and animal biodiversity. Read more in our June 3 news release.

Signup for General CRP ran from March 4 to 29, 2024. General CRP includes a Climate-Smart Practice Incentive to help increase carbon sequestration and reduce greenhouse gas emissions by helping producers and landowners establish trees and permanent grasses, enhance wildlife habitat and restore wetlands. Read more in our March 2 news release.

FSA is now accepting applications for the reopened Continuous CRP signup. FSA encourages agricultural producers and landowners interested in conservation opportunities for their land in exchange for yearly rental payments to consider the enrollment options available. Read more in our January 11, 2024 news release.

USDA is encouraging agricultural producers and private landowners enrolled in CRP to learn more about the opportunities with FSA's Forest Management Initiative (FMI). Through the initiative, USDA provides financial assistance to those with active CRP contracts with forest cover to improve the health of their forests. To be eligible, you must have properly completed authorized forest management practice activities to improve the condition of resources, promote forest management, and enhance wildlife habitat. Read more in our January 25, 2024 news release.

2023 Enrollment

In 2023, producers nationwide enrolled 927,000 acres through the General CRP Signup, more than 2.3 million acres through Grassland CRP Signup, and

694,000 acres through the Continuous CRP Signup. These 2023 signups bring the current total of acres involved in CRP enrollments to 24.8 million.

Also, FSA issued more than $1.77 billion in October 2023 to the more-than 667,000 agricultural producers and landowners for more than 23 million acres enrolled in CRP in 2023. Read more in our October 16, 2023 news release.

Improvements to CRP and Climate Incentives and Results

Since 2021, FSA has taken steps to improve the program. This included higher payment rates, new incentives, and a more targeted focus on CRP's role in climate change mitigation. FSA introduced a new climate-smart practice incentive for General CRP and Continuous CRP designed to reward participants who implement conservation practices that increase carbon sequestration and reduce greenhouse gas emissions. To learn more about updates to CRP, read our "What's New with CRP" fact sheet.

Meanwhile, FSA also launched a CRP Climate Change Mitigation Assessment Initiative to collaborate with three assessment teams to monitor the impact of CRP perennial grasses, tree plantings and wetlands on soil carbon and greenhouse gases. In addition, USDA and other researchers are conducting field experiments to improve cover establishment as the climate changes.

FSA also renewed its emphasis on CREP. This includes hiring additional staff, working to improve the flexibility of the program, and increasing participation among tribes. In 2022, FSA entered into its first-ever CREP agreements with Tribal Nations—reflecting priorities and goals of USDA to broaden the scope and reach of its voluntary, incentive-based conservation programs to engage underserved communities, the Cheyenne River, Oglala and Rosebud Sioux Tribes entered into CREP agreements with FSA to enroll eligible grassland, pastureland, and other agricultural lands within the boundaries of their reservations in this conservation program.

CRP and Disaster Assistance

CRP haying and grazing provides for emergency haying and grazing on certain CRP practices in a county designated as D2 or higher on the U.S.

Drought Monitor, or in a county where there is at least 40 percent loss in forage production. Our haying and grazing webpage resources include eligible county maps, tables and CRP practices. For producers not in an eligible county, there are options available under non-emergency haying and grazing provisions outside of the primary nesting season.

Chapter 2

Conservation Reserve Program (CRP)*

U.S. Department of Agriculture

Overview

USDA Farm Service Agency's (FSA) Conservation Reserve Program (CRP) is a voluntary program that contracts with agricultural producers so environmentally sensitive agricultural land is not farmed or ranched, but instead devoted to conservation benefits. CRP participants establish long-term, resource- conserving plant species, such as approved grasses or trees (known as "covers") to control soil erosion, improve water quality and develop wildlife habitat. In return, FSA provides participants with rental payments and cost-share assistance. Contract duration is between 10 and 15 years.

CRP is authorized by the Food Security Act of 1985 and was reauthorized by the Agricultural Improvement Act of 2018 (the 2018 Farm Bill). The 2018 Farm Bill was extended by the Further Continuing Appropriations and Other Extensions Act, 2024 (Pub. L. 118-22). The program is also governed by regulations published in 7 CFR, part 1410. The program is implemented by FSA on behalf of USDA's Commodity Credit Corporation.

Program Administration

FSA administers CRP while technical support functions are provided by: USDA's Natural Resources Conservation Service (NRCS); State forestry

* This is an edited, reformatted and augmented version of U.S. Department of Agriculture, Farm Service Agency Publication, dated March, 2024.

In: The Conservation Reserve Program
Editor: Jerry P. Norton
ISBN: 979-8-89530-774-8

agencies; Local soil and water conservation districts and; non- federal technical service providers.

Benefits

CRP protects more than 20 million acres of American topsoil from erosion and is designed to safeguard the nation's natural resources. By reducing water runoff and sedimentation, CRP protects groundwater and helps improve the condition of lakes, rivers, ponds and streams. CRP is also a major contributor to increased wildlife populations in many parts of the country.

General CRP Enrollment

Producers have the opportunity to offer land for general CRP enrollment annually during announced enrollment periods. Offers for CRP contracts are ranked according to the Environmental Benefits Index (EBI). FSA uses the following EBI factors to assess the environmental benefits for the land offered:

- Wildlife habitat benefits resulting from covers on contract acres;
- Water quality benefits from reduced erosion, runoff and leaching;
- On-farm benefits from reduced erosion;
- Benefits that will likely endure beyond the contract period;
- Air quality benefits from reduced wind erosion and;
- Cost.

Further information on general CRP signup is available in the FSA fact sheet "Conservation Reserve Program General Signup."

Continuous CRP Enrollment

Under continuous CRP enrollment, environmentally sensitive land devoted to certain conservation practices may be enrolled in CRP at any time.

Certain eligibility requirements still apply, but offers are not subject to competitive bidding. Further information on continuous CRP enrollments is available in the FSA fact sheet "Conservation Reserve Program Continuous Enrollment."

Grasslands Enrollment

Grasslands CRP helps landowners and operators protect grassland, including rangeland, and pastureland, and certain other lands, while maintaining the areas as grazing lands. The program emphasizes support for grazing operations, plant and animal biodiversity, and grassland and land containing shrubs and forbs under the greatest threat of conversion. Further information on Grassland CRP enrollments is available in the FSA fact sheet "Conservation Reserve Program Grassland."

For More Info

This fact sheet is provided for informational purposes only; other restrictions may apply.

For details, contact your local FSA office. To find your local FSA office, contact your local service center and USDA Farm Service Agency office: farmers.gov/service-locator.

Who is Eligible?

A producer must have owned or operated the land for at least 12 months prior to submitting the offer for continuous or 12 months before the close of general or grassland signup, unless:

- The new owner acquired the land due to the previous owner's death;
- The ownership change occurred due to foreclosure where the owner exercised a timely right of redemption in accordance with state law or;
- The circumstances of the acquisition present adequate assurance to FSA that the new owner did not acquire the land for the purpose of placing it in CRP.

Eligible Land

Both continuous CRP and general CRP require cropland, land to be planted or considered planted to an agricultural commodity for four of six crop years from 2012 to 2017, and be physically and legally capable of being planted (no planting restrictions due to an easement or other legally binding instrument) in a normal manner to an agricultural commodity.

For general signup, land also must meet one of the following criteria:

- Have a weighted average erosion index of eight or higher;
- Be enrolled in a CRP contract that expires Sept. 30 or;
- Be located in a national or state conservation CRP priority area. Marginal pasture land may also be eligible for continuous signup.

Grasslands CRP has alternative requirements for eligible land.

Payments

FSA provides CRP participants with annual rental payments, as well as certain incentive payments and costshare assistance.

In return for establishing long-term, resource-conserving covers, FSA provides annual rental payments to participants. FSA bases rental rates on the relative productivity of the soils within each county and the average cash rent using data provided by the National Agricultural Statistics Service (NASS). The soil rental rates are subject to an 85 percent proration for general signup and a 90 percent proration for continuous signup. The maximum CRP rental rate for each offer is calculated in advance of enrollment. Producers may offer land at that rate or offer a lower rental rate. CRP grasslands rates are based on 75 percent of the NASS pasture rate.

Chapter 3

Conservation Reserve Enhancement Program: Fiscal Year 2023*

U.S. Department of Agriculture

Executive Summary

Introduction

The Conservation Reserve Enhancement Program (CREP) is a part of the Conservation Reserve Program (CRP) under which the U.S. Department of Agriculture (USDA), through the Farm Service Agency (FSA), and non-Federal partners co-invest in partner-led projects which target regional and national resource concerns.

CREP was administratively established in February 1997 through a final rule in 7 CFR 1410.50 authorizing USDA to enter into agreements with States to address resource concerns. By October 1997, USDA executed its first CREP agreement with the State of Maryland targeting acreage that, once enrolled, reduces runoff, sediment, and nutrients into the Chesapeake Bay and promotes enhanced wildlife habitats. The 2018 Farm Bill established CREP as a permanent subprogram under CRP and codified program requirements.

Through CREP agreements, financial, educational, and technical assistance is provided to producers who voluntarily implement certain conservation measures on land in lieu of continued agricultural production. Each agreement identifies the specific geographic area and conservation

---

* This is an edited, reformatted and augmented version of U.S. Department of Agriculture Report to the United States Congress.

In: The Conservation Reserve Program
Editor: Jerry P. Norton
ISBN: 979-8-89530-774-8

concerns to be addressed and the commitments of both USDA and the partner. Partners are required to submit annual reports regarding the status of each project.

This report to Congress is provided in accordance with Section 1231A of the Food Security Act of 1985, as amended, which requires the Secretary of Agriculture to submit an annual report to Congress that describes, with respect to each agreement, the: 1) status of the agreement; 2) purpose and objectives of the agreement; 3) Federal and eligible partner commitments made under the agreement; and 4) progress made in fulfilling those commitments.

Program Improvements and Investments in Fiscal Year (FY) 2023

FSA has rolled out several CREP improvements and strategic investments of time and resources to promote and support Secretary Vilsack's priority for USDA to work with agriculture, forestry, and rural communities to develop Climate Smart Agriculture and Forestry practices and solutions. These include:

- *Engage New Partners.* CREP continues to engage with new partners to support the success of new and upcoming agreements, such as the three new CREP agreements with Tribal governments and the newly minted Big Sioux River Watershed agreement with the state of South Dakota, which has seen its first enrollment in FY 2023 and expects further enrollments in FY 2024. FSA continues to explore partnership opportunities with states, Tribes, and non-governmental organizations. By engaging more partners, FSA can expand the reach of CREP and co-invest in conservation-focused efforts with groups that provide a direct link to historically underserved producers and landowners.
- *Provide Flexibility to Meet Local Needs.* CREP is one of the most flexible tools FSA has for locally driven, partner-led conservation efforts to reward producers and drive important environmental and climate outcomes. FSA continues to find ways to use these flexibilities to shape CRP to address the conservation priorities of State and national significance most important to partners, such as increased grassland productivity, mitigation of local drought concerns, and remediation of frequently flooded areas. The FSA CREP team

continues to engage with state partners to provide technical assistance to achieve these goals.

- *Streamline Administrative Processes Associated with New and Existing CREP Agreements.* Over the past fiscal year, FSA has taken several steps to improve communication with partners, streamline the reporting process, and improve the quality of information reported. These efforts have resulted in more consistent and effective partner reporting. In FY 2023, FSA entered into an interagency agreement with conservation experts at the U.S. Geological Survey (USGS) to adopt new software and processes to streamline progress reporting and aid in compiling the report to Congress, as well as proposing new metrics for tracking conservation benefits based on current science.

Additionally, FSA has built on the success of the CREP agreement templates from FY 2022 by streamlining the process of drafting new, and revising existing, CREP agreements. For example, during FY 2023, FSA completed a revision to the Colorado Republican River CREP to add dryland agricultural use provisions to further address regional drought concerns and encourage permanent retirement of water rights. FSA also approved four new CREP agreements (the South Dakota Cheyenne River Sioux Tribe CREP, the South Dakota Oglala Sioux Tribe CREP, the South Dakota Rosebud Sioux Tribe CREP, and the South Dakota Big Sioux River Watershed CREP).

FY 2023 Success Stories

- *Habitat Conservation and Creation.* Creation and conservation of wildlife habitat, particularly for vulnerable species and species of local importance, remains a key goal of many CREPs. For example, the South Dakota-James River CREP reported that they were able to more than quadruple public fishing and hunting acreage in Hutchinson County, SD to over 8,000 acres, which will provide key wildlife habitat as well as an economic benefit to small communities in the area. This CREP has also facilitated the enrollment of 58,000 acres of cropland in Brown, Marshall, Spink, and Day counties. These counties have experienced two major flood events in the past decade. These enrollments are expected to yield water quality and flood reduction benefits, as well as create large blocks of habitat for grassland nesting birds.

- *Improved Water Quality.* Water quality is targeted by 33 of the 38 active CREP projects. The Indiana CREP focuses on the Conservation Practice (CP)31 Bottomland Timber Establishment practice as a means of protecting water quality, with nearly 100,000 additional trees planted across 167 acres in FY 2023 and a total of 4.3 million trees planted across Indiana watersheds since the CREP's inception in 2005. Indiana has made significant investments in the CREP program since FY 2022, with the State Soil Conservation Board approving $2 million, more than 3 times the amount appropriated in prior years, to make improvements which include, but are not limited to, expanding the CREP project area to include additional watersheds in the state. These improvements are expected to reduce the amount of sediment, phosphorus, and nitrogen entering rivers and streams in targeted watersheds. In a similar vein, the Maryland CREP has also been using trees to advance water quality goals. In 2021, the Maryland CREP received state partner support in the form of an increased signing bonus for CP22 Riparian Buffer practice enrollments and authority to hire 13 additional staff members to support recent state legislation to plant and maintain an additional 5 million trees by 2030, including 500,000 trees in underserved communities. Nine of the 13 staff members approved for the program have been hired, and the Maryland legislature has increased the state signing bonus from $100 to $1,000 per acre for all new CP- 22 enrollments.
- *Strong Engagement and Outreach.* Outreach to farmers, landowners, and other key stakeholders is a major focus for many CREPs, especially those with new or renewed agreements. Partners in South Dakota's Big Sioux River Watershed noted a successful kickoff year, having enrolled the first seven contracts in the agreement in FY 2023 in partnership with Pheasants Forever Farm Bill Biologists. In Ohio's Lake Erie Watershed, CREP partners credit a new CREP agreement with dramatically improving communication between all partners, which has led to improved implementation. Strong engagement with participants has also led to high contract reenrollment in some areas, such as 87 percent reenrollment in South Dakota's James River Watershed and over 50 percent reenrollment in Nebraska's Platte Republican Resource Area.

Goals and Next Steps

- Continue to Improve Reporting. Each year, an emphasis is placed on improving the CREP report by advancing the quality of reporting and streamlining the process for state partners and USDA staff. For the FY 2024 report, FSA will continue to work with subject matter experts at USGS to review reporting tools to streamline the process, reduce the potential for human error, and provide additional conservation metrics for adoption by interested states. FSA will also continue to engage with CREP partners through hosting webinars and providing technical assistance to state partners in order to accurately and comprehensively capture program benefits.
- Identify Additional Conservation Opportunities that can be Addressed through CREP. FSA continues to explore ways in which CREP can be used to address new and existing conservation issues in a state or throughout the nation. These may include issues of acute local importance such as resource scarcity and environmental degradation. The structure of CREP provides partners the flexibility to emphasize local conservation priorities in the future.
- Focus on Outcome-Driven Results. As CREP engages new partners and continues to strive for greater program participation, program results must be outcome-driven to be impactful. As new CREP agreements are negotiated and existing agreements are revised, a greater emphasis will be placed on selecting and measuring outcomes that contribute to FSA mission success, benefiting both farmers and the environment. This process will be conducted in consultation with the USGS team to ensure that the metrics selected are consistent with current best practices in conservation science. The process will also incorporate results from other ongoing FSA research projects dealing with ecosystems services from CREP and related programs and the policy issues surrounding said programs.[1]

[1] For examples of ongoing research conducted under FSA's Monitoring, Assessment, and Evaluation (MAE) program, please consult this link.

CREP Program Summary

Since CREP was authorized in 1997, USDA has entered into 58 CREP agreements with state partners, 38 of which are active. As of September 30, 2023, over 697,500 acres were under contract, including acres that remain enrolled under CREP agreements that are no longer active.[2] The South Dakota James River CREP (79,537.0 acres), Pennsylvania Chesapeake Bay Watershed CREP (71,138.4 acres), and the Ohio Scioto River Watershed CREP (59,146.0 acres) have the highest total enrollment; these three CREPs account for 30 percent of total CREP enrollment.

FY 2023 CREP Program Summary

In FY 2023, 61,309 acres were enrolled through 29 separate agreements in 24 states. Of those 61,309 acres, 20,802 acres, or approximately 34 percent, were new enrollments in FY 2023. The South Dakota James River CREP (4,906.7 acres), the Idaho Eastern Snake River Basin CREP (3,059.9 acres), and the Ohio III-Scioto River Watershed CREP (2,075.0 acres) accounted for 48 percent of all new enrollments in FY 2023. At the close of FY 2022, 40,522. acres in expiring contracts were re-enrolled into new FY 2023 contracts. The South Dakota James River CREP (12,343.9 acres), the Ohio III-Scioto River Watershed CREP (6,566.7 acres), and the Pennsylvania Chesapeake Bay Watershed CREP (6,337.1 acres) made up 62.3 percent of all re- enrollments in FY 2023.

Federal contributions totaled $162,683,849 in FY 2023 for annual rental, signing incentive, practice incentive, and cost-share payments for acreage enrolled in active CREP agreements. CREP partners reported $66,270,049 in cash and in-kind contributions for FY 2023.

---

[2] An active CREP agreement refers to any agreement that is currently in effect and has the potential to accept enrollment.

A successful establishment in the New York III Statewide CREP.

FY 2023 Status of Agreements

In FY 2023, eligible new or re-enrolling acreage was enrolled via 29 separate agreements in 24 states. There were nine additional active agreements under which no acres were enrolled. For additional information regarding specific agreements without FY 2023 enrollment, please see the footnotes on this and the following page. There were also a number of agreements that are no longer active due to termination or suspension, but which have active acres enrolled under a previous signup. As reported in the "Agreement Goals and Progress" section of this report, many of the active agreements with low or no enrolled acres are conducting ongoing outreach efforts aimed at potential CREP participants, often in tandem with state and local partners.

State	*CREP Agreement*	*New Acres in FY 2023*	*Re-enrolled Acres in FY 2023*	*Total acres as of Sept. 30, 2023*
Arkansas*	Arkansas I-Bayou Meto			28.5
Arkansas*	Arkansas II-Cache River/Bayou DeView			390.0
Arkansas*	Arkansas III-Illinois River			11.1
Colorado	Colorado I-Republican River	1,387.9	0.0	11,699.0
Colorado*	Colorado II-High Plains			699.4
Colorado[3]	Colorado III-Rio Grande	0.0	0.0	10,024.0
Delaware	Delaware	1.2	31.3	2,865.3
Hawaii*	Hawaii			1,268.0
Idaho	Idaho-Eastern Snake River Plain	3,059.9	213.1	10,226.7
Illinois	Illinois	486.4	79.7	48,104.7
Indiana	Indiana	466.4	1,263.0	21,683.6
Iowa	Iowa	93.9	0.0	3,302.2
Kansas[4]	Kansas-Upper Arkansas River	0.0	0.0	22,691.3
Kentucky*	Kentucky-Green River			15,952.5
Maryland	Maryland	613.7	2,445.3	40,513.5
Michigan	Michigan	423.3	963.9	18,895.5
Minnesota*	Minnesota II - Mississippi			57.5
Minnesota*	Minnesota II - Missouri/Des Moines			151.2
Minnesota*	Minnesota II - Red River			41.0
Minnesota	Minnesota Water Quality	563.3	341.7	30,274.1
Mississippi	Mississippi Delta	333.2	0.0	6,022.6
Missouri*	Missouri			3,068.1
Montana	Montana - Missouri and Madison River	0.0	231.7	10,484.0
Montana[5]	Montana - Upper Clark Fork River Basin	0.0	0.0	0.0
Nebraska*	Nebraska I-Central Basins Resources			1,183.2
Nebraska	Nebraska II-Platte Republican Resource Area	823.1	641.1	40,047.0
New Jersey	New Jersey	33.4	42.7	556.5
New York	New York I-New York City	10.7	59.5	1,243.0

[3] The Colorado III-Rio Grande CREP Agreement has been in effect since FY 2013. The partner reported decreased enrollment which is related to recent drought conditions and producer interest in other annual fallowing and drought relief programs. The partner and State FSA staff continue to work on outreach to encourage enrollment.

[4] The Kansas CREP Agreement was revised in August 2021 to expand the CREP project area effective June 2022. Over 400 acres were enrolled in FY 2023 but are not included in this report because the contracts do not take effect until FY 2024.

[5] The Montana - Upper Clark Fork River Basin CREP signup opened in FY 2021. The partner and State FSA staff continue to work on outreach and education to encourage enrollment.

State	*CREP Agreement*	*New Acres in FY 2023*	*Re-enrolled Acres in FY 2023*	*Total acres as of Sept. 30, 2023*
New York[6]	New York II-Syracuse	0.0	0.0	43.8
New York	New York III-Statewide	189.1	28.1	5,904.5
North Carolina	North Carolina	34.6	0.0	3,165.9
North Dakota[7]	North Dakota Riparian Project	0.0	0.0	1,051.2
Ohio	Ohio I-Lake Erie	1,122.7	1,639.3	50,162.2
Ohio*	Ohio II-Upper Big Walnut Creek			9.3
Ohio	Ohio III-Scioto River Watershed	2,075.0	6,566.7	59,146.0
Oklahoma*	Oklahoma			491.1
Oregon	Oregon	730.6	3,110.1	44,252.8
Pennsylvania	Pennsylvania I-Chesapeake Bay Watershed	1,614.1	6,337.1	71,138.4
Pennsylvania	Pennsylvania II-Ohio River Basin	115.8	1,558.8	10,098.0
Pennsylvania[8]	Pennsylvania III - Delaware River Basin	0.0	0.0	90.3
South Dakota	South Dakota- Big Sioux River	784.4	0.0	784.4
South Dakota	South Dakota-James River	4,906.7	12,343.9	79,537.0
South Dakota[9]	Cheyenne River Sioux Tribe	0.0	0.0	0.0
South Dakota[10]	Oglala Sioux Tribe	0.0	0.0	0.0
South Dakota[11]	Rosebud Sioux Tribe	0.0	0.0	0.0
Vermont	Vermont	58.0	203.7	2,015.5
Virginia	Virginia I-Southern Rivers	124.5	221.5	5,932.0
Virginia	Virginia II-Chesapeake Bay	25.9	541.6	8,933.9
Washington	Washington	14.5	345.9	13,135.3

[6] The New York II-Syracuse CREP Agreement has been in effect since FY 2002. The partner and State FSA staff continue to work on outreach and education to encourage enrollment.

[7] The North Dakota Riparian Project has been in effect since FY 2017. A grant used by the State for this CREP project expired in September 2023, greatly reducing the state funding available for this project.

[8] The Pennsylvania III-Delaware River Basin CREP Agreement has been in effect since FY 2017. The partner and State FSA staff continue to focus on outreach to encourage enrollment in the program.

[9] The Cheyenne River Sioux Tribe CREP Agreement was approved on March 1, 2023. Over 5,600 acres were enrolled in FY 2023 but are not included in this report because the contracts do not take effect until FY 2024.

[10] The Oglala Sioux Tribe CREP Agreement was approved on October 9, 2022. Over 3,000 acres were enrolled in FY 2023 but are not included in this report because the contracts do not take effect until FY 2024.

[11] The Rosebud Sioux tribe CREP Agreement was approved on October 20, 2022. While acreage was not enrolled in FY 2023, additional outreach and education is planned to encourage participation.

State	*CREP Agreement*	*New Acres in FY 2023*	*Re-enrolled Acres in FY 2023*	*Total acres as of Sept. 30, 2023*
West Virginia	West Virginia	106.2	452.0	4,662.1
Wisconsin	Wisconsin	603.4	845.5	35,558.6
TOTAL		20,801.9	40,507.4	697,595.6

* Terminated or suspended CREP agreement with active contract acreage enrolled under a previous CREP signup.

Fy 2023 Federal And Eligible Partner Commitments

The USDA issued a total of $162,683,849 in FY 2023 for all active agreements. Partner contributions totaled $66,270,049. Federal and partner contributions by agreement are provided below.

State	Agreement Name	FY 2023 Cash and in Kind Commitments			
		Federal Cash	Partner Cash	Partner In-Kind*	Total Partner
Colorado	Colorado I-Republican River	$2,405,610	$182,299	$199,500	$381,799
Colorado	Colorado III-Rio Grande	$1,776,097	$415,712	$6,002,573	$6,418,285
Delaware	Delaware	$469,278	$134,176	$99,131	$233,307
Idaho	Idaho-Eastern Snake River Plain	$2,168,805	$190,008	$9,440,386	$9,630,394
Illinois	Illinois	$14,437,944	$7,974	$1,529,830	$1,537,804
Indiana	Indiana	$6,111,196	$2,062,500	$285,900	$2,348,400
Iowa	Iowa	$1,702,221	$616,954	$0	$616,954
Kansas	Kansas-Upper Arkansas River	$2,987,697	$149,044	$281,545	$430,589
Maryland	Maryland	$10,447,848	$2,571,597	$482,206	$3,053,803
Michigan	Michigan	$3,679,146	$682,299	$545,500	$1,227,799
Minnesota	Minnesota Water Quality	$10,421,519	$10,728,048	$1,977,076	$12,705,124
Mississippi	Mississippi Delta	$803,502	$28,989	$12,500	$41,489
Montana	Montana - Missouri and Madison River	$671,133	$0	$0	$0
Montana	MT - Upper Clark Fork River Basin	$0	$0	$11,077	$11,077
Nebraska	Nebraska II- Platte Republican Resource Area	$7,638,731	$3,117,367	$3,117,367	$6,234,734
New Jersey	New Jersey	$554,090	$53,319	$85,203	$138,522
New York	New York I-New York City	$328,097	$201,811	$201,881	$403,692
New York	New York II- Syracuse	$5,565	$500	$500	$1,000

State	Agreement Name	FY 2023 Cash and in Kind Commitments			
		Federal Cash	Partner Cash	Partner In-Kind*	Total Partner
New York	New York III-Statewide	$1,083,077	$7,043,427	$0	$7,043,427
North Carolina	North Carolina	$455,650	$926,559	$19,417	$945,976
North Dakota	North Dakota Riparian Project	$55,133	$45,719	$2,816	$48,535
Ohio	Ohio I-Lake Erie	$15,071,400	$2,796,802	$235,500	$3,032,302
Ohio	Ohio III-Scioto River Watershed	$20,632,630	$78,994	$14,000	$92,994
Oregon	Oregon	$4,622,976	$852,373	$1,103,179	$1,955,552
Pennsylvania	Pennsylvania I-Chesapeake Bay Watershed	$12,295,483	$79,890	$395,265	$475,155
Pennsylvania	Pennsylvania II- Ohio River Basin	$1,328,882	$5,980	$0	$5,980
Pennsylvania	Pennsylvania III - Delaware River Basin	$17,898	$336	$0	$336
South Dakota	South Dakota- Big Sioux River	$31,748	$36,803	$11,071	$47,874
South Dakota	South Dakota- James River	$10,101,501	$3,492,146	$40,400	$3,532,546
Vermont	Vermont	$524,563	$95,374	$0	$95,374
Virginia	Virginia I- Southern Rivers	$983,734	$346,748	$21,782	$368,530
Virginia	Virginia II- Chesapeake Bay	$1,060,263	$26,429	$0	$26,429
Washington	Washington	$3,667,758	$2,235,446	$0	$2,235,446
West Virginia	West Virginia	$749,473	$3,425	$8,828	$12,253
Wisconsin	Wisconsin	$8,613,550	$431,958	$444,852	$876,810
TOTAL		$147,904,200	$39,641,006	$26,569,285	$66,210,291

* Partner in-kind activities may include outreach and promotion, monitoring, program administration, and related activities.

A CP22 riparian buffer planting in the New York I – New York City CREP.

Agreement Purpose

Agreement Name	Agreement Purpose
Colorado I-Republican River	Reduce the amount of irrigation water consumptive use, conserve energy, and reduce agricultural chemicals and sediment from entering waters of the State from agricultural lands. Enhance aquatic and terrestrial wildlife habitat through establishment of permanent vegetative cover.
Colorado III-Rio Grande	Improve water quantity and quality, enhance habitat for wildlife populations, reduce irrigation water consumptive use, and reduce agricultural chemical and sediment runoff within the Rio Grande Basin in Colorado.
Delaware	Enhance water quality through reduction of agricultural nutrients to further the goal of restoring designated uses of the State's waterbodies and to enhance wildlife habitats.
Idaho-Eastern Snake River Plain	Improve water quantity and quality, enhance wildlife habitat through establishment of vegetative cover, reduce irrigation water consumptive use, and reduce agricultural chemical and sediment runoff to the waters of the State.
Indiana	Improve water quality and address wildlife issues by reducing erosion, sedimentation and nutrients, and enhancing wildlife habitats.

Agreement Name	Agreement Purpose
Illinois	Reduce the amount of agricultural chemicals and sediment from agricultural lands entering waters within the project area. In addition, enhance aquatic and terrestrial wildlife habitat through establishment of permanent vegetative cover.
Iowa	Improve water quality by reducing nitrate loads to surface waters, enhance wildlife habitat, and increase recreational opportunities.
Kansas-Upper Arkansas River (UAR)	Reduce the amount of irrigation water consumptive use and reduce agricultural chemicals and sediment from entering waters of the State from agricultural lands that contribute to poor water quality in rivers and alluvial and high plains aquifers.
Maryland	Help reduce the occurrence of runoff, sediment, and nutrient accumulation in the Chesapeake Bay and promote enhanced wildlife habitats.
Michigan	Reduce sediment, phosphorus, and nitrogen in the surface water supply of the Macatawa, River Raisin, Saginaw Bay watersheds and the Michigan portion of Western Lake Erie Basin Watershed. Improve water quality, enhance habitat for fish and wildlife, and enhance nesting for upland birds, mammals, and waterfowl.
Minnesota Water Quality	Improve water quality in the project area by establishing and maintaining buffers of permanent vegetation between eligible waterbodies and adjacent agricultural cropland, restoring and maintaining wetlands; establish permanent vegetative cover in wellhead protection areas, and establish beneficial habitat for terrestrial and aquatic wildlife habitat.
Mississippi Delta	Assist in the maintenance and/or improvement of current water quality conditions through the reduction of agricultural sources of sediment, nutrients, and waterborne pathogens in the targeted watershed.
Montana-Missouri/Madison River	Improve water quality and enhance fish and wildlife habitat along the Missouri and Madison Rivers.
Montana- Upper Clark Fork River Basin	Address water quality issues by the reduction of sediment, nutrients and other pollutants entering rivers and streams, restoring and maintaining wetlands, and the enhancement of habitat within the Upper Clark Fork River Basin.
Nebraska II-Platte Republican Resource Area	Improve water quantity and quality, and the enhancement of wildlife habitat, through establishment of vegetative cover to reduce irrigation water consumptive use and agricultural chemical and sediment runoff into the waters of Nebraska.
New Jersey	Improve water quality, reduce impairment from agricultural non-point sources, and restore the ecological functions of New Jersey streams.
New York I- New York City	Enhance water quality by reduction of agricultural sources of sediment, nutrients, and waterborne pathogens and the enhancement of wildlife habitats.
New York II-Syracuse	Assist in the maintenance of current water quality conditions through the reduction of agricultural sources of sediment, nutrients, and waterborne pathogens.

Agreement Name	Agreement Purpose
New York III-Statewide	Maintain or improve current water quality conditions through the reduction of agricultural sources of sediment, nutrients, and waterborne pathogens.
North Carolina	Enhance water quality by the reduction of sediment and nutrients, and the enhancement of fisheries and wildlife habitats for State and federally listed threatened and endangered species.
North Dakota Riparian Project	Improve water quality by establishing and maintaining buffers of permanent vegetation between the waterbody and adjacent agricultural cropland, and to provide benefits to pollinators, including honeybees, and other wildlife by establishing beneficial habitat and forage
Ohio I-Lake Erie	Reduce the amount of sediment and nutrients entering the targeted watersheds from agricultural sources.
Ohio III-Scioto River Watershed	Improve water quality, reduce agricultural nonpoint source pollution to surface waters, and reduce soil erosion.
Oregon	Assist in the recovery of fish species which have been listed as threatened or endangered species under the Federal Endangered Species Act.
Pennsylvania I-Chesapeake Bay Watershed	Reduce nutrient and sediment loading of the Upper and Lower Susquehanna and Potomac River Basins, improving water quality, enhancing wildlife, and producing nutrient reductions.
Pennsylvania II-Ohio River Basin	Reduce nutrient and sediment loading of the Ohio, Monongahela, and Allegheny Rivers and the smaller streams and tributaries in the same watersheds.
Pennsylvania III-Delaware River Basin	Reduce nutrient and sediment loading of the Delaware River and the smaller streams and tributaries in the same watersheds.
South Dakota- Big Sioux River	Improve water quality, reduce soil erosion, enhance wildlife habitat, and create public hunting and fishing access.
South Dakota-James River	Improve water quality, reduce soil erosion, improve flood control, enhance water wildlife habitat, and create public hunting and fishing access.
Vermont	Achieve non-point source pollutant reduction, enhance fish and wildlife habitat, and attain conservation goals established by the State.
Virginia I-Southern Rivers	Enhance water quality by reduction of sediment and nutrients within the Southern Rivers Basin and the enhancement of fish and wildlife habitats for State and federally listed threatened and endangered species.
Virginia II-Chesapeake Bay	Reduce the amount of nutrients entering estuaries from agricultural sources to assist Virginia in achieving its nutrient reduction goals for agriculture in the area, significantly reduce the amount of sediment entering water courses, and enhance habitat for the preservation of natural diversity of biological resources.
Washington	Assist in the recovery of salmon species that have been listed as threatened or endangered species under the Federal Endangered Species Act.
West Virginia	Reduce the runoff, sediment, and nutrients from agricultural enterprises into the boundaries of the designated watersheds, which will promote improved water quality and enhanced wildlife habitats.

Agreement Name	Agreement Purpose
Wisconsin	Improve the water quality of several waterbodies that drain agricultural lands throughout the State of Wisconsin, through a reduction of sediment and the nutrient loading to these waterbodies.

A West Virginia CREP riparian buffer zone.

Agreement Goals And Progress (As Reported by Partner)

Agreement Name	Goal Type	Goal	Progress and Methods Used
Colorado I-Republican River	Water Quantity	Reduce irrigation water use for agricultural purposes from the Ogallala Aquifer by six and one- half percent (6.5%) from 2004 levels.	• Average irrigation consumption in the Republican River CREP is approximately 1.5 acre-foot of irrigation water per acre of land irrigated. Water reduction is calculated by multiplying the total number of irrigated acres retired through CREP since 2005 times 1.5, resulting in an estimated reduction of 50,118 acre-feet of irrigation water per year. • Irrigation use has been reduced by 7.4% percent to date, which exceeds the goal.
	Water Quantity	Reduce irrigation water consumption by 49,800 to 69,200 acre-feet annually.	• Measuring devices have been required on all wells within the district since 2009. Well pumping is reported annually and is highly variable being dependent on climate each year. Therefore, actual well pumping is not used to measure progress toward the water quantity goal. • Using the metric of 1.5 acre feet per acre (af/ac), which is a standard average value over time recognized for this basin, irrigation water use has been reduced by 50,118 ac/ft per year (considering acres enrolled since 2005, including expired contracts) exceeding the minimum goal of 49,800 af per year.
	Soil Erosion	Reduce soil erosion from 751,633 tons to approximately 259,395 tons per year, a total reduction of approximately 492,238 tons per year.	• Erosion reduction is based on the original goal of 492,238 tons/ac for 35,000 acres of the Republican River CREP. • Since this CREP requires permanent retirement of the irrigation component, it is

Agreement Name	Goal Type	Goal	Progress and Methods Used
			appropriate to use the total number of acres enrolled over time, including those where contracts have expired.
			• Multiplying the tons/ac by the 33,412 irrigated acres enrolled since 2005 indicates that 469,903 tons have been reduced and the goal is 95 percent achieved.
	Water Quality	Reduce annual fertilizer and pesticide application by a minimum of 3,865 tons per year from 2004 levels.	• NRCS and Colorado State University provided an annual application rate for fertilizers and pesticides for acres under irrigation within the CREP area. The calculation reduces that number to zero for each acre enrolled in CREP. • For the 33,412 acres, the reduction in application is 3,865 tons/year. Thus, the goal is approximately 95 percent achieved.
	Energy Reduction	Reduce the total annual use of electricity by 2.76 million kilowatt hours (kwh) through reductions in groundwater pumping.	• Each center pivot is operated by electricity and monitored by several power companies within the region. The electricity is quantified by acre and the total number of kwh/yr is calculated by multiplying the 32,034 acres retired since 2005 through CREP by the electricity it takes to irrigate the enrolled acres. • This equates to 2,634,769 kwh/yr. The goal is approximately 95 percent achieved.
		Enroll up to 2,500 eligible cropland acres in CP100, Annual Crop Production, Non-irrigated.	• Dryland cropping provisions and practice CP100, • Annual Crop Production, Non-irrigated were added to the CREP agreement effective June 1, • 2023. No acres were enrolled in practice CP100 in FY 2023.

Agreement Name	Goal Type	Goal	Progress and Methods Used
Colorado III-Rio Grande	Water Quality and Soil Health	Reduce soil erosion from 681,252 tons to 149,487 tons per year on all acres enrolled in CREP, a total reduction of 531,765 tons.	• Soil erosion is determined by soil type and is quantified by NRCS for irrigated agriculture and permanent cover. Progress is measured by calculating the difference between the two and multiplying it by the number of acres enrolled. • Approximately 136,000 tons/year have been reduced. Twenty-six percent of this goal has been achieved to date.
	Habitat	Establish up to 40,000 acres of habitat for numerous wildlife species, including several aquatic and wetland dependent species that are declining due to habitat loss and degradation.	• Approximately 25 percent of the 40,000 acres have been enrolled to date and due to the arid environment and complex aquifers, habitat goals will not be realized for many years.
	Water Quality	Reduce annual fertilizer and pesticide application from enrolled acres by approximately 3,650 tons per year from 2012 levels.	• Progress is measured by using a baseline of fertilizer and pesticide application rates per acre and multiplying by the total number of acres enrolled. • Almost 26 percent of this goal has been met to date for a reduction of approximately 948 tons per year.
	Water Quality and Habitat	Establish up to 40,000 acres of native vegetation throughout the project area.	• More than 25 percent of the 40,000 total acres • (10,024 acres) have been enrolled to date. • All enrolled acres are being checked by NRCS and Subdistrict #1 staff to ensure native vegetation is being established on enrolled acres.
	Water Quality and Habitat	Restore and enhance up to 750 acres of degraded temporary and permanent wetlands.	• Up to 750 acres are targeted for this goal. However, there have been no signups and no interest expressed for this practice.
	Water Quantity	Reduce irrigation water use for agricultural purposes of the confined and unconfined aquifer within the targeted watershed by approximately 60,060 acre-feet of ground	• Approximately 25 percent (10,024 acres) have been enrolled in this CREP to date. Annual irrigation water reduced is approximately 15,763 acre-feet.

Agreement Name	Goal Type	Goal	Progress and Methods Used
		water per year, equal to almost 12 percent water savings within the project area and 5 percent water savings throughout the entire Rio Grande basin in Colorado.	
	Water Quantity	Increase stream flows in streams associated with the watershed within the project area.	• Streamflows are monitored regularly by the Colorado Division of Water Resources and Subdistrict #1; however, other factors, such as the extreme flashiness of the hydrograph, snowpack in surrounding mountains, precipitation variability during the growing season, multi-year groundwater transit times, large scale recharge operations, and other factors make it impossible to attribute any specific changes to CREP enrollment in this very large basin. • Even so, reduction in consumption on 10,509 acres increases available water for streamflows as that water is either not diverted from the stream or remains in the groundwater system where it can migrate to streams.
	Energy Reduction	Reduce energy consumption from an average- sized (125 acre) pivot from approximately 47,000 kilo-watt hour (kW-hr) per year to less than 14,000 kW-hr per year for the first three years during cover establishment on all pivots enrolled in the CREP. Subsequent years' energy consumption will be reduced to zero for all pivots enrolled in the CREP.	• Each 125-acre pivot enrolled uses the data in the goal to estimate the total energy savings by year. • With 84 equivalent pivots enrolled, the total reduction to date is approximately 30 million kW-hrs this year and 80 million KW-hrs to date or 38 percent of the potential maximum of this goal over 15 years.
	Water Quality	Reduce the percentage of groundwater test wells	• Baseline and current nitrogen in groundwater information was generated from the

Agreement Name	Goal Type	Goal	Progress and Methods Used
		containing nitrogen (NO3) levels above EPA standards.	Colorado Department of Agriculture and Colorado State University resources for the Upper Rio Grande. • There are 15 wells with data pre- and post-CREP in the 200,000-acre subdistrict that were reviewed. The EPA drinking water standard is 10 ppm. The natural level of nitrate is 2.5 ppm for this area. For the 15 wells, the average nitrate level went down slightly to 6.28 ppm while the median went up slightly to 3.81 ppm. Further review of the sampling information indicates that 2 of the 5 sampling wells that had levels exceeding the EPA standard in 2011 had levels below the EPA standard in the latest round of sampling.
Delaware	Water Quality	Facilitate nutrient and sediment reduction pursuant to the goal of restoring designated uses of surface waters.	• Models can estimate cumulative CREP nutrient reductions by watershed for nitrogen (N), phosphorous (P) and sediment. The CREP practices in Delaware contribute to improving the State's water quality. By having the following CREP practices in place from FY 2008- 2023, the estimated amount of each nutrient pollutant load that will not reach surface and/or ground waters on a yearly basis include: CP 3A – 1,872.13 acres, 88,732.56 of N and 2,368.07 lbs. of P reduction; CP4D – 346.85 acres, 13,281.88 lbs. of N and 429 lbs. of P reduction; CP9 – • 158.69 acres, 6,437.18 of N and 127.62 lbs. of P reduction; CP21 – 266.1 acres, 9,797.37 lbs. of N and 296.88

Agreement Name	Goal Type	Goal	Progress and Methods Used
			lbs. of P reduction; CP22 – 52.02 acres, 2,354.89 lbs. of N and 73.15 lbs. of P Reduction; CP23 - 89.2 acres, 3,048.7 lbs. of N and 141 lbs. of P reduction; CP23A – 80.52 acres, 2,744.75 lbs. of N and 127.88 lbs. of P reduction. The total active acres under CREP practices from 2008–2023 is 2,865.3 acres with a cumulative total of N and P estimated load reductions of 126,397.33 lbs. and 3,563.60 lbs., respectively. • Load reductions are routinely incorporated into the Total Maximum Daily Load (TMDL) development process both for crediting purposes and for future reductions from additional best management practice (BMP) implementation.
	Habitat	Increase wildlife habitat and create wildlife corridors in the Chesapeake Bay, Delaware Bay, and Delaware Inland Bays.	• Delaware set a goal of establishing 10,000 acres of selected practices to meet the goals of the CREP Program. The Cumulative CREP Program Activity from FY2007-2023 is 2,865.3 acres which are currently under 10 to 15-year contract terms. This number accounts for contract loss and renewal. For
			• Signup 57 re-enrollment (contracts scheduled to expire 9/30/2022), there was a loss of 89.9 acres due to three (3) Producers not wanting to re-enroll their conservation practices and six (6) Producers having compliance issues which resulted in those practices not being re-enrolled. Five (5) of the Producers contracts were in the Chesapeake Bay

Agreement Name	Goal Type	Goal	Progress and Methods Used
			Watershed and the other four (4) contracts were in the Delaware Bay watershed. There was one (1) new Producer contract for 1.2 acres located in the Chesapeake Bay Watershed. There were 9 re-enrolled State contracts written for Signup57. The new and re-enrolled contracts by conservation practices for the three (3) watersheds are as follows: • (4) Delaware Bay, (6) Chesapeake Bay, and (1) Delaware Inland Bay. This represents 38.57 acres of new and/or renewal practices for Signup57. These acres are represented by the following conservation practices; 7.1 acres of CP3A; 8.4 acres of CP4D, 7.8 acres of CP9, 5.9 acres of CP21; and 3.3 acres of CP22. • Each of these practices are acres that were formerly farmland or marginal pastureland. By installing and managing these conservation practices, it provides wildlife habitat that will be protected from disturbances. As these types of conservation practices are implemented, progress has been made to restore wildlife populations by increasing desirable habitat areas. Delaware has currently • achieved approximately 28.66% of the 10,000-acre conservation practice implementation goal.
	Water Quality	Restore natural conditions for water temperature and dissolved oxygen in areas protected by riparian forested buffers.	• The Delaware Department of Natural Resources and Environmental Control (DNREC) Watershed

Agreement Name	Goal Type	Goal	Progress and Methods Used
			Assessment and Management Section conducts
			• routine water quality monitoring throughout the state of Delaware at designated monitoring stations. Every two years, the state of Delaware combines watershed assessment reports, and the Watershed Assessment and Management Section publishes a report focusing on comprehensive water quality monitoring. The NRCS Guidelines for CREP practice establishment criteria for CP22 Riparian Forested Buffers state that they must be adjoining a Category 1 or 2 stream, ditch, water body or non-farmed wetland, or adjoin drainage ditches contributing to Category 1 and 2 impaired segments. In Delaware, this practice is a Riparian Forested Buffer, which is predominantly trees. This practice provides shade for adjacent water bodies which helps control the water temperature and dissolved oxygen levels. • There are 52.02 acres of riparian buffers that are planted along Delaware's waterways that are currently enrolled in CREP. • The DNREC Division of Watershed Stewardship conducted stream assessments throughout the state to survey for freshwater mussels. The freshwater mussel survey program conducts presence/absence surveys. The presence and/or increased presence of freshwater mussels are commonly used as

Agreement Name	Goal Type	Goal	Progress and Methods Used
			an indicator species for water quality. Four surveys were conducted between June and August 2023 using a new ArcGIS Field Maps application to collect the data. On June 9, 2023, DNREC survey crew members completed a freshwater mussel presence/absence study of White Clay Creek in the Christina Basin watershed in New Castle County, DE and no mussels were observed. On June 15, 2023, the team surveyed Brecknock County Park – Isaac
			Branch in the St. Jones River watershed in Kent County, DE. They recorded 10 individuals, all Eastern Elliptio (E. complanate) ranging in size from 7.5-13 cm. On July 6, 2023, DNREC surveyed Fork Branch Nature Preserve, Cahoon Branch, in the St. Jones River watershed in Kent County, DE. Thirteen mussel individuals were examined and were all found to be Eastern Elliptio and ranged in size from 5.5-8.5 cm. The final survey that was conducted took place on August 30, 2023, at Chapel Branch in the Nanticoke River watershed in Sussex County, DE. No freshwater mussels were observed. • In the future, quantitative results from stream surveys are desired with the goal of measuring mussel population increases and decreases relating to water quality and habitat.
	Water Quality	Provide conservation buffers on approximately 1,200 miles	

Agreement Name	Goal Type	Goal	Progress and Methods Used
		of the State's waterways and drainage systems.	• In FY 2023, cumulative practice acres total 2,706.61 acres (2,865.3 Total acres – 158.69 CP9 acres) which include the following practices: CP3A-Hardwood Trees (1,872.13 acres), CP4D-Permanent Wildlife Habitat (346.85 acres), CP21- Filter Strips, (266.1 acres) CP22-Riparian Forested Buffers (52.02 acres) and CP23/23A-Wetland Restoration (169.72 acres). These conservation practices all must have a side along a waterway or drainage area. The only practice not included in the list above is the CP9 – Shallow Water Areas for Wildlife. This practice does not need to be installed along a stream or body of water, so the CP9 acres are being left out of the acreage calculations. • To estimate the length of conservation buffers, an approach was used measuring the square root of a 43,560 square foot acre, which represents the side of the square and is ~ 209 feet or 0.04 miles (=4/100 of a mile). Since there are 2,706.82 acres along Delaware waterways and drainage systems, they can be multiplied by the 209 feet to equal 565,725.38 feet. To convert 565,725.38 feet to miles, divide by 5,280 feet per mile which equals 107.14 miles. Currently, there are approximately 107.14 miles out of the 1,200 miles covered with conservation buffers from CREP in Delaware's waterways and drainage systems. Approximately 8.93 percent of the State's

Agreement Name	Goal Type	Goal	Progress and Methods Used
			waterways and drainage systems have one of the CREP conservation buffers alongside.
Idaho-Eastern Snake River Plain	Water Quantity	Enroll up to 50,000 acres of eligible cropland.	• 10,226.7 acres of cropland are enrolled.
	Water Quantity	Cease the application of water on up to 50,000 acres of irrigated cropland and thereby reduce consumptive use of water from the Eastern Snake Plain Aquifer by an estimated 2 acre-feet annually per enrolled acre up to a total of 100,000 acre-feet annually.	• 20,533.4 acre/ft annually of groundwater is conserved by the 10,226.7 acres of farmland enrolled in the CREP program in Idaho.
	Habitat	Provide up to 50,000 additional acres of native grassland habitat for wildlife in the Idaho CREP project area, improving habitat for grassland- nesting birds such as sage grouse and sharp-tailed grouse.	• Idaho currently has 10,214.04 acres of native and mixed native-introduced grassland habitat enrolled under this agreement. • Improving water quantity helps with flow and temperature for aquatic species. Upland wildlife and bird habitat benefit from permanent vegetative cover and protection during nesting seasons. • Upland wildlife and bird habitat benefits from permanent vegetative cover and protection during nesting seasons. CREP in Idaho has used exclusions during upland bird nesting seasons to limit disturbance during vulnerable nesting times.
			• Columbian sharp-tailed grouse are one of six subspecies of sharp-tailed grouse. Currently, there are approximately 56,000 to 61,500 Columbian Sharp-tailed Grouse across their range and Idaho supports about 62.5 percent of the overall U.S. population. This information is cited in the

Agreement Name	Goal Type	Goal	Progress and Methods Used
			latest population estimate as published in Idaho Fish & Game "2015 – 2025 Columbian Sharp-tailed Grouse Management Plan for Idaho." • Throughout their historical range, the decline of Columbian sharp-tailed grouse has been primarily attributed to the loss of native grassland and shrub- grassland vegetation types. • CRP and CREP have generally had a major positive impact on sharp-tail/upland bird populations in Idaho. Between these programs, hundreds of thousands of acres were seeded to a mixture of perennial grasses and forbs. • The extensive, undisturbed grassland that developed on these seeded areas has provided secure nesting and brood-rearing habitat, especially where alfalfa/forbs was included in the seeding mix. • Currently, approximately 70 percent of Columbian sharp-tailed grouse habitat occurs on private land in Idaho, with the remaining 30 percent on public land. In most areas, the birds are dependent on both private and public land to meet their seasonal habitat requirements.
Illinois	Water Quality	Decrease sediment, nitrogen, and phosphorus loading by 10 percent within stream reaches associated with land enrolled in CRP through this CREP, when compared to 1998 pre-enrollment averages.	• To assist with evaluating the progress towards meeting the Illinois CREP goals, a scalable load estimation modeling framework developed at the Illinois State Water Survey (ISWS) was used to set up load estimation models for the 12 Hydrologic Unit Code (HUC)-12

Agreement Name	Goal Type	Goal	Progress and Methods Used
			watersheds that cover the monitored CREP watersheds in the Illinois River and the Kaskaskia River watersheds. The annual percent reduction of sediment and nutrient loads were estimated by comparing the loads simulated using installed CREP practices in the watershed with that of a no-CREP scenario in any given year. The estimated loads were bias-corrected using observed sediment and nutrient data collected by ISWS. Water year 2022 data (the most recently available data) was used for the calculations for this report. • For the Illinois River Watershed: • The total water year 2022 CREP practice acreage in the monitored HUC-12s of the Illinois River Watershed accounted for 2.1 percent of the watershed area and resulted in estimated load reductions of 5,109 tons of sediment (2.54 percent), 28,659 lbs. of nitrogen (1.31 percent), and 4,207 lbs. of phosphorus (1.22 percent). • The total load reductions since 2000 were estimated to be 87,295 tons of sediment (2.23 percent), 705,522 lbs. of nitrogen (1.37 percent), and 109,310 lbs. of phosphorus (1.47 percent). • For the Kaskaskia River Watershed: • The total water year 2022 CREP practice acreage in the monitored HUC-12s of the Kaskaskia River Watershed was only 0.2 percent of the watershed area and the

Agreement Name	Goal Type	Goal	Progress and Methods Used
			resulting load reductions were 221 tons of sediment (0.25 percent), 1,073
			• lbs. of nitrogen (0.11 percent), and 283 lbs. of phosphorus (0.11 percent). • Since water year 2015 (the first year of data available for analysis for the Kaskaskia River watershed), the total load reductions in the monitored HUC-12s were estimated to be 1,990 tons of sediment (0.26 percent), 12,680 lbs. of nitrogen (0.15 percent) and 3,781 lbs. of phosphorus (0.15 percent). • Illinois CREP is making progress towards meeting the 10 percent reduction in sediment, nitrogen, and phosphorus goal. As noted above, only 2.1 percent of the Illinois River Watershed and 0.26 percent in Kaskaskia River Watershed are enrolled in CREP, but a reduction is still shown in sediment, nitrogen, and phosphorus. Percent load reductions vary from HUC-12 to HUC-12 based on CREP enrollments. Of all monitored HUC-12s, the largest total load reduction in 23 years (water year 2000-2022) was obtained for 7130005080-Haw Creek in the Illinois River Watershed, which also has the highest CREP acreage enrollment of 6.5 percent. This resulted in percent load reductions of 7.39 percent for sediment, 5.45 percent for nitrogen and 6.19 percent for phosphorous. To determine the progress towards meeting the CREP goals, eligible CREP areas

Agreement Name	Goal Type	Goal	Progress and Methods Used
			were determined for each of the monitored watersheds and in 2022, a total of 175 acres of land were eligible for CREP in the monitored HUC-12s. The largest area of eligible CREP land (40.7 acres) was obtained for 71300050803-Haw Creek, indicating the possibility of additional load reduction with future enrollments in this watershed.
	Habitat	Increase the population of avian "Species of Greatest Conservation Need," as defined in the Illinois Wildlife Action Plan, on land enrolled in CRP through this CREP by 50 percent, when compared to similar land managed for agricultural production.	• Illinois Natural History Survey researchers conducted 182 multi-species point counts at 111 randomly selected points in 81 CREP easements within the Illinois and Kaskaskia River watersheds. • The combined density of all avian species was 5 times greater on CREP compared to similar land managed for agriculture. For avian "Species of Greatest Conservation Need" (SGCN), increases in density on CREP compared to crop fields ranged from 84 - 5,616 percent, exceeding the goal to increase densities by 50 percent. • Additionally, we recorded 22 SGCN using CREP fields compared to a 1990 study that recorded 7 SGCN using agricultural fields in Illinois, and a 2014 study that only recorded 3.
	Habitat	Increase the average number of native fish species and environmentally sensitive aquatic insects (stoneflies, caddisflies, and mayflies) by 15 percent within stream reaches associated with land enrolled in CRP through this CREP when compared to	• The Illinois Natural History Survey has continued efforts to monitor the impacts of CREP on aquatic life in Illinois. A total of 18 wadeable stream sites were sampled under base-flow conditions to evaluate fish assemblages, benthic

Agreement Name	Goal Type	Goal	Progress and Methods Used
		similar streams surrounded by land managed for agricultural production. The stream reaches sampled for this goal rotate between streams in the Kaskaskia Basin and the Illinois Basin, annually.	macroinvertebrate assemblages, stream habitat, and water chemistry in the Kaskaskia River Basin. • In the Kaskaskia River Basin, CREP reaches have 44 percent more native fish species present on average than the agricultural reaches, and 57 percent of streams met or exceeded the 15 percent goal. CREP reaches in the Illinois River Basin have 7 percent fewer native fish species present on average than agricultural reaches, but 27 percent of streams meet or exceed the 15 percent goal and 45 percent had more species present. • In the Kaskaskia River Basin, none of the sampled streams had more sensitive aquatic insect species present in CREP reaches than agricultural reaches,
			whereas 73 percent of streams in the Illinois River basin had at least 15 percent more environmentally sensitive insect species present in CREP reaches. • Sensitive aquatic insect species made up a higher percentage of the total aquatic insect community in CREP reaches than in agricultural reaches in 57 percent and 46 percent of streams in the Kaskaskia and Illinois river basins, respectively. • More monitoring data will be needed to estimate the biological effect of CREP and determine the statistical significance of the estimation.
Indiana	Water Quality	Protect a minimum of 3,000 linear miles of watercourses	• The Indiana State Department of Agriculture (ISDA) tracks

Agreement Name	Goal Type	Goal	Progress and Methods Used
		through the installation of conservation buffer practices.	all practices that are enrolled and that have signed a state landowner participation agreement form. When a project is recorded as completed in the state's tracking system, the CREP Leader records the length in feet of the practice that is protecting a body of water. (The definition for a body of water is taken from the CRP Manual.) The feet are then converted to miles. • Through the installation of conservation buffer practices in the Indiana CREP, approximately 900 linear miles of watercourses are currently protected within the Indiana CREP watersheds. Overall, this is 30 percent of the goal to protect 3,000 linear miles of watercourses in the targeted CREP watersheds.
	Water Quality	Reduce the amount of sediment, phosphorus, and nitrogen entering rivers and streams in the designated watersheds by 2,450 tons per year of sediment, 2,400 pounds per year of phosphorus, and 4,700 pounds per year of nitrogen.	• ISDA uses the Region 5 Sediment and Nutrient Load Reduction Model developed by the Environmental Protection Agency (EPA) to estimate the sediment, nitrogen and phosphorus load reductions from individual best management practices installed on the ground. CREP Leaders apply this model to each conservation practice enrolled and installed through
			• the CREP to estimate the positive effects of the practice on water quality. • Practices installed in FY 2023 resulted in an approximate reduction of 6,518 tons of sediment, 8,193 lbs. of phosphorus, and 16,290 lbs. of nitrogen. Overall, since the

Agreement Name	Goal Type	Goal	Progress and Methods Used
			expansion of the program in 2010, approximate reductions include 92,445 tons of sediment, 108,446 lbs. of phosphorus, and 213,317 lbs. of nitrogen.
	Water Quality	Increase the acres of wetlands in the watersheds for erosion control, sediment reduction, stormwater retention, and nutrient uptake.	• In FY 2023, according to the state's tracking system, approximately 700.9 acres of wetland restorations were completed/installed on the ground, and 677.6 acres were enrolled.
	Water Quality	Seek enrollment of up to 26,250 acres of eligible cropland, including frequently flooded agricultural lands and restorable wetlands.	• According to ISDA's tracking system, there are approximately 22,545 acres that have been enrolled in the Indiana CREP over the life of the program since 2005, with approximately 21,453 of those acres completed. There are 21,683 acres that are under contract in FY 2023. • The number of acres that the state shows as enrolled and completed is different than what FSA has under contract because of the differences in when practices are recorded by each system. FSA records acres as soon as the contract is approved and shows as active in the federal system, while the state records acres once the practice under which they are enrolled has been implemented.
Iowa	Water Quality	Add the capacity to remove 300 to 600 tons of nitrate-nitrogen per year from agricultural tile drainage. Over the next ten years, this would add wetland capacity to annually remove 3,000 to 6,000 tons of nitrate-nitrogen from agricultural tile drainage. Over a 60-year life, each wetland acre	• Two wetlands were constructed in 2023 which remove an estimated 14.85 tons of nitrogen from receiving waters. These wetlands will remove 2,228 tons of nitrogen over their 150-year lifespan. • Cumulatively, all the CREP wetlands constructed to date remove 823 tons per year of

Agreement Name	Goal Type	Goal	Progress and Methods Used
			nitrogen at a weighted average cost of $495 per ton of nitrogen removed from the receiving waters.
		would remove approximately 20 to 40 tons of nitrate-nitrogen.	• In support of the Iowa CREP, Iowa State University monitors a representative subset of wetlands each year and performs mass balance analyses to document nitrate reduction. By design, the wetlands selected for monitoring span a wide range in wetland-to-watershed area ratio and inflow nitrate concentrations. This ensures that monitoring results represent a broad range of factors most affecting wetland performance including hydraulic loading rate, residence time, nitrate concentration, and nitrate loading rate. In addition to documenting wetland performance, the ongoing monitoring and research programs allow continued refinement of modeling and analytical tools used in site selection, design, and management of CREP wetlands. • Twelve wetlands were monitored in 2023, including 11 Iowa CREP wetlands and one mitigation wetland (DD15-N). • Monitoring and Analysis: • Wetland monitoring included measurements of wetland inflows, outflows, pool elevations, water temperature, and collection of weekly to biweekly water quality grab samples and daily samples. Most sites were equipped for

Agreement Name	Goal Type	Goal	Progress and Methods Used
			flow monitoring and automated water sampling • Some wetlands have compound weir structures allowing for discharge calculations based on water elevation and calibrated weir equations. Where these weirs were not available, wetland inflow and/or outflow channels were instrumented with submerged area velocity • (SAV) Doppler flow meters and stage recorders for close-interval (every five minutes) measurement of flow velocity and stream depth. The SAV measurements were combined with cross-sectional channel profiles and stream depth to calculate discharge as the product of water velocity and wetted cross-sectional area.
	Water Quality	Reduce sediment entering surface waters in the Lake Panorama Watershed by 80,000 tons per year following filter strip and riparian buffer establishment. This has the potential of significantly reducing total sediment accumulation into Lake Panorama.	• Progress toward meeting Goal 2 has been very limited. While some landowners have installed grade control structures or filter strips using state cost- share programs in recent years, no landowners have signed up for the available CREP incentives in the watershed. • Additional staff have been hired to support the implementation of edge-of-field practices statewide • and will be utilized to support implementing these practices within the Lake Panorama Watershed.
	Water Quality	Reduce or maintain soil erosion on agricultural land enrolled in the CREP to below the soil loss tolerance level for the soils present (2-5 tons per acre).	• The wetland upland buffer seeding is required to be a CP-23 practice consisting of perennial native warm season grasses and native forbs. This perennial vegetation protects the soil and keeps soil erosion

Agreement Name	Goal Type	Goal	Progress and Methods Used
			to a minimum. Soil erosion estimates made for these sites document that annual soil erosion rates are less than one ton per acre per year. • Additionally, CREP sites are dominated by low- gradient slopes. If the upland buffer area includes steeper slopes, these areas are generally small acreages and are also protected by the perennial vegetation and plant residues.
	Water Quality	Demonstrate a variety of available wetland technologies and their value for improving water quality.	• Two types of wetland development technologies are being researched, evaluated, and demonstrated currently. The first technique is "Tile Zone Wetlands"; the second is wetland development on floodplain landscapes.
Kansas-Upper Arkansas River	Water Quantity	Enroll a maximum of 40,000 acres into CRP through this CREP in the project area (to be comprised of up to 37,000 irrigated cropland acres and up to 3,000 acres of dryland center-pivot corners).	• 2022. Over 400 acres were enrolled in FY 2023 but are not included in this report because the contracts do not take effect until FY 2024.
	Water Quantity	Reduce the consumption of groundwater for irrigation in the project area by 65,000 acre-feet of permitted water appropriations based on the enrollment of 37,000 irrigated acres.	• Progress is measured by recording the amount of annual water appropriation rights that are voluntarily retired in conjunction with CRP enrollment of irrigated lands within the project area. • . The cumulative project total as of 09/30/2023 is 47,643.7 acre-feet of annual water appropriation water rights retired. The 2 CREP enrollments approved in FY23 will add an additional 1,040 acre- feet of annual water savings, effective in FY24. • The goal of having 65,000 acre-feet of annual water rights appropriations retired is on track at 73.2 percent.

Agreement Name	Goal Type	Goal	Progress and Methods Used
	Soil Conser-vation	Reduce agricultural use of land that is unsuitable for dryland farming by enrolling 25,000 acres with a Wind Erodibility Index (WEI) of 134 or greater.	• Progress is measured by recording CRP enrollments in the UAR CREP which are classified as being in soil types of WEI of 134 or greater. • The project total as of 09/30/2023 is 21,304.7 acres enrolled in soil types of WEI of 134 or greater. The 2 CREP enrollments approved in FY23 will add an additional 436.73 acres of soil types of WEI of 134 or greater, effective in FY24. • The goal of having 25,000 acres being enrolled in Tier 1 Unsuitable soils is on track at 85.2 percent.
	Soil Conser-vation	Reduce the amount of annual soil lost to erosion by approximately 150,000 tons per year.	• Progress is measured by recording CRP enrollments in the UAR CREP and multiplying the total acres enrolled by a 'tons of soil saved' soil erosion conservation factor of 4 tons/acre/year. • Two additional CRP enrollments of 426.73 acres were recorded in FY 2023, which added 1,746.92 tons of soil saved to the total cumulative project enrollment to date for this goal.
			• The project total for this goal is 90,765.32 tons of soil saved. The two additional CRP enrollments of 426.73 acres recorded in FY 2023 will add an additional 1,746.92 tons of soil saved to the total cumulative project enrollment, effective FY24. • FY23 project acres represent 56.7% of the project acreage allotment of 40,000 acres. When looking at this goal proportionately to the enrollment acreage, this goal

Agreement Name	Goal Type	Goal	Progress and Methods Used
			would equate to 85,050 annual tons of soil saved. Based on the 22,691.3 acres of enrollment, 90,764 annual tons of soil are saved from erosion. This goal is being met at 106.7 percent when analyzed in this manner. Overall, the goal is 60.5 percent met.
	Energy Conser-vation	Reduce the total annual use of electricity by 16 million kilowatt hours when full enrollment is met.	• Progress is measured by recording CRP enrollments in the UAR CREP and multiplying the estimated number of center pivots by a "total energy savings per center pivot" conservation factor of 59,850 kilowatt hours per center pivot. • The project total for this goal is now 10,832,850 kilowatt hours saved. The two additional CREP enrollments of 426.73 acres recorded in FY 2023 (4 center pivots) will add 239,400 kilowatt hours saved to the total cumulative project enrollment for this goal, effective in FY24. • This goal is on track at 67.7 percent of the total kilowatt hours saved.
	Agricul-tural Transition	Demonstrate the ability to gradually transition from irrigation farming to non-irrigation farming through enrollment of 1,000 acres of land that has a Wind Erodibility Index of less than 134 and will be devoted to CRP practice CP43.	• Progress is measured by recording the number of • CRP acres enrolled in conservation practice CP43 - "Prairie Strips". • This goal was added to the CREP authorizing • Memorandum of Agreement on August 9, 2021. • No offers for this conservation practice have yet to be received.
Maryland	Water Quality	Reduction of nutrient pollution from agricultural lands of approximately 11,500,000 pounds of	• Maryland's Watershed Implementation Plan (WIP) outlines specific actions and strategies that it will take to

Agreement Name	Goal Type	Goal	Progress and Methods Used
		nitrogen and 1,100,000 pounds of phosphorus on an annual basis.	achieve pollution limits set by EPA by 2025. Maryland listed 41 agricultural milestone actions in their goals to accelerate Chesapeake Bay restoration and meet the total maximum daily load (TMDL) goal. These actions included four categories of CREP practices: highly erodible land protection, riparian grass buffers, riparian forest buffers, and wetlands. • Progress toward each practice is tracked and reported though the Soil Conservation Districts and reported annually to EPA. Subsequently, a suite of modeling tools is employed by the EPA Chesapeake Bay Program to estimate nutrient and sediment delivery across the watershed. In addition to newly installed practices, the Maryland Department of Agriculture (MDA) completes site inspections and field verification over the lifespan of the BMPs to ensure that water quality benefits remain. • Based on enrolled acreage through FY 2023, progress is estimated at 1,732,736 pounds of nitrogen and 31,872 pounds of phosphorus annually reduced through CREP.
	Water Quality	Reduction of sediment loading into streams from agricultural lands of approximately 200,000 tons of sediment on an annual basis.	• Progress for FY 2023 is estimated at 25,969, tons of sediment reduced annually.
	Habitat	Increase the survivability, distribution, and abundance of targeted fish, wildlife, and plant species in the Chesapeake Bay region. Emphasis on Rare and	• CREP acres have expanded conservation cover for critical wildlife habitat. Of the 40,513.5 acres enrolled as of 9/30/2023, 16,335 (40 percent) were within the BioNet

Agreement Name	Goal Type	Goal	Progress and Methods Used
		Declining Species Habitat enrollments will be given to high priority species listed under the North American Waterfowl Management Plan and State and Federally-listed threatened or endangered species, such as bald eagles, Delmarva fox squirrel, Eastern bog turtle, dwarf wedge mussel, glassy darter, and harparella.	mapping of important areas for rare species.
			• For tree planting practices alone (CP22, CP3, CP4D, CP3A), current CREP projects in Maryland generated over $16 million in ecosystem services annually. These ecosystem services include wildlife habitat, nitrogen removal, groundwater recharge, flood protection, surface water treatment, and air quality improvement. In addition, the importance of the program to birds in need of conservation can be estimated using available data, updated for the current acres enrolled. CREP practices provided critical breeding habitat to 91,989 birds, an increase of 74,741 additional birds compared to non-CREP acres. • Grassland and shrubland birds—including Species of Greatest Conservation Need such as northern bobwhite, grasshopper sparrows, eastern meadowlarks, and prairie warblers— have benefitted greatly, with an additional 50,238 birds using habitat created via CREP buffer practices vs. other areas. • Forest buffers provide habitat for an estimated 26,372 birds.

Agreement Name	Goal Type	Goal	Progress and Methods Used
			• CREP wetland practices have provided valuable habitat for waterfowl species such as the regionally-prioritized American black duck, with an estimated 2,262,615 days of duck use resulting from current CREP enrollments.
Michigan	Water Quality	When full enrollment (80,000 acres) is reached, annually reduce sediment entering rivers and streams by 52,000 tons.	• With 18,895.5 acres enrolled in the CREP program as of Sept. 30, 2023, the annual amount of sediment entering rivers and streams is being reduced by 20,875 tons. • 40 percent of this goal has been met.
	Water Quality	When full enrollment is reached annually reduce phosphorus entering rivers and streams by 52,000 pounds.	• With 18,895.5 acres enrolled in the CREP program as of Sept. 30, 2023, the annual amount of phosphorus entering rivers and streams is being reduced by 30,438 pounds. • 58 percent of this goal has been met.
	Water Quality	When full enrollment is reached annually reduce nitrogen entering rivers and streams by 105,000 pounds.	• With 18,895.5 acres enrolled in the CREP program as of Sept. 30, 2023, the annual amount of nitrogen entering rivers and streams is being reduced by 92,740 pounds. • 88 percent of this goal has been met.
	Soil Health	When full enrollment is reached annually sequester approximately 60,000 metric tons of carbon.	• With 18,895.5 acres enrolled in the CREP program as of Sept. 30, 2023, the annual CO_2 equivalent of carbon being sequestered is 17,595 metric tons. • 29 percent of this goal has been met.
Minnesota Water Quality	Water Quality	Reduce phosphorus by 19,200 pounds per year.	• MN CREP is designed to improve water quality in the project area by establishing and maintaining buffers of permanent vegetation between eligible waterbodies and adjacent agricultural cropland,

Agreement Name	Goal Type	Goal	Progress and Methods Used
			restoring and maintaining wetlands, establishing permanent vegetative cover in wellhead protection areas, and establishing beneficial habitat for terrestrial and aquatic wildlife. • While the original reduction estimates were to the field scale, a sub-watershed (12-digit hydrologic unit code outlet; 12-digit HUC outlet) approach is used here due to the restoration of marginal cropland enrolled in the MN CREP often taking years to become a fully functioning site, depending on the conservation practice type. The State of Minnesota has completed Hydrologic Simulation Program Fortran (HSPF) watershed modeling for most of the major watersheds in the State. The Minnesota • Pollution Control Agency (MPCA) provided the Minnesota Board of Water and Soil Resources
			• (BWSR) with average pollution reduction coefficients in pounds/acre of phosphorus for equivalent conservation practices by individual 12-digit HUC watersheds. BWSR calculated pollution reduction estimates by multiplying the acres treated by the coefficients from specific conservation practice types. • For phosphorus reductions, the sub-watershed scale approach should provide comparable results to the field scale due to the nature of fate and transport of phosphorus within the

Agreement Name	Goal Type	Goal	Progress and Methods Used
			HSPF model. Note that watershed outlet values will have slightly lower pollution reductions compared to field scale methods. • The BWSR used the most current nutrient reduction coefficients to calculate nutrient reductions. The reduction of phosphorus pollutant loading is 13,261 pounds per year for new practices installed through MN CREP. Existing watercourse and wetland practice acres were excluded from this calculation because these existing features would not likely be converted back to cropland. Existing grassland practices are included in the calculations. Tree/shrub practices, existing and new, are excluded because the information needed to calculate the nutrient reductions is not available in the HSPF model.
	Water Quality	Reduce nitrogen by 1,220,000 pounds per year.	• The approach used above is also used here. The reduction of nitrogen pollutant loading is 434,164 pounds per year for new practices installed through MN CREP. Existing watercourse and wetland practice acres were excluded from this calculation because these existing features would not likely be converted back to cropland. Existing grassland practices are included in the calculations. Tree/shrub practices, existing and new, are excluded because the information needed to calculate the nutrient reductions is not available in the HSPF model.

Agreement Name	Goal Type	Goal	Progress and Methods Used
	Water Quality	Reduce sediment runoff by 123,000 tons per year.	• Again, the approach used above is used here. For sediment reductions, the sub-watershed scale approach provides less comparable results to the field scale due to the HSPF model's more limited ability to process complex in-stream processes for sediment fate, transport, and attenuation from the headwaters of a watershed to the watershed outlet. In the case of sediment and total suspended solids, the sub-watershed scale pollution reductions from the HSPF model are significantly less than calculating field scale loads with conventional soil loss models such as the Revised Universal Soil Loss Equation, Version 2 (RUSEL2) or the Water Erosion Prediction Project (WEPP) model. • The BWSR used the most current nutrient reduction coefficients to calculate nutrient reductions. The reduction of nitrogen pollutant loading is 1,659 tons per year for new practices installed through MN CREP. Existing watercourse and wetland practice acres were excluded from this calculation because these existing features would not likely be converted back to cropland. Existing grassland practices are included in the calculations. Tree/shrub practices, existing and new, are excluded because the information needed to calculate the nutrient reductions is not available in the HSPF model that is used.

Agreement Name	Goal Type	Goal	Progress and Methods Used
Mississippi Delta	Habitat	Increase native bottomland forest acreage, which is considered to be one of the most degraded ecosystems in North America and provides habitat for several migratory waterfowl and shorebird species as well as the federally threatened Louisiana black bear.	• While not specified in the CREP agreement, to accomplish this goal, the partner will restore 8,000 acres of bottomland hardwoods and plant 3,480,000 trees within the Mississippi Delta. • As of Sept. 30, 2023, 6,022.6 acres have been enrolled connecting fragmented forest and creating travel corridors for expanding wildlife populations in the Mississippi Delta region while providing critical habitats. This represents 75.3 percent of the 8,000- acre goal.
	Water Quality	Reduce sediment loading in streams and lakes in the project area by installing conservation measures which reduce erosion rates and reduce off-field transportation rates of herbicides, pesticides, and nutrients. This would lead to water quality improvements benefiting the federally endangered Sheepnose mussel (Plethobasus cyphyus capax) and the Pallid sturgeon (Scaphirhynchus albus).	• The amount of excess sediment, nitrogen, and phosphorous reduced and calculated on an acre per year basis is used to determine the total volume of these nutrients leaving the fields and entering the waterways. • Using the RUSLE model, predicted per acre annual reduction rates are 10,822 lbs. of sediment, 5.4 lbs. of nitrogen, and 10.8 lbs. of phosphorous.
	Habitat	Restore wetlands with CRP practice CP23, in a coordinated effort by Federal and State agencies and non-governmental organizations. Assist producers in establishing an increased acreage amount on Mississippi-adapted bottomland hardwood native tree species to serve as habitat for several migratory waterfowl and shorebird	• CP23 is implemented on marginal agricultural lands that frequently flood. • As of Sept. 30, 2023, a total of 797.42 acres were enrolled in CP23, Wetland Restoration.

Agreement Name	Goal Type	Goal	Progress and Methods Used
		species, as well as the federally threatened Louisiana black bear.	
	Water Quality	Provide increased protection of sub-surface water sources from contamination by agricultural chemicals, nutrients, and pathogens by assisting with the installation of conservation measures that reduce point and non-point pollution. Assist producers in establishing increased riparian buffer acreage to serve as nutrient/chemical uptake and filtering sites.	• Land enrolled in all CREP practices reduce point and non-point source pollution by conserving sediment, nitrogen, and phosphorous and preventing these contaminants from entering the Yazoo River Watershed. • As of Sept. 30, 2023, 6022.6 total acres have been enrolled in this CREP project.
Montana-Missouri and Madison River	Habitat	Establish, restore, and improve up to 11,000 acres of riparian buffers along approximately 524 miles of the Missouri and Madison Rivers and their tributaries within a two-mile corridor in nine counties. This will be the key practice, given the importance of riparian habitat. The specific measurable item will be acres of riparian habitat restored. Riparian habitat restoration will help recover candidate species and species of special concern such as the arctic grayling and westslope cutthroat trout and listed Threatened and Endangered species such as the pallid sturgeon.	• The goal is being measured by the number of acres that are established, restored, or improved within the two-mile corridor of the Missouri or Madison Rivers. • There were no new enrollments this year, but 231.7 acres were re-enrolled in FY 2023, and the agreement had 10,484 total acres enrolled at the end of the fiscal year.
	Habitat	Establish and improve up to 14,000 acres of permanent, native, and introduced wildlife habitat within the 2-mile-wide corridor of the Missouri and Madison Rivers. Many wildlife species depend on both riparian and upland habitats. Native and introduced grassland and	• The specific measurable criteria of success will be acres of permanent wildlife habitat restored. • There was no additional progress this year. There are currently no acres enrolled in CP-4D (permanent wildlife habitat) under this agreement.

Agreement Name	Goal Type	Goal	Progress and Methods Used
		scrubland habitat establishment will support recovery of these species as well as a number of ground-nesting grassland obligates. In addition, establishment of permanent native and introduced cover on highly erodible cropland will support the water quality and fishery goals of this agreement by reducing sediment inputs to water bodies. The specific measurable criteria of success will be acres of permanent wildlife habitat restored.	
	Water Quality & Habitat	Restore up to 1,000 acres of degraded wetlands within a 2-mile-wide corridor of the Missouri and Madison Rivers. Wetland restoration within the corridor will support the fish and wildlife habitat restoration and water quality goals of the project.	• The specific measurable criteria of success will be acres of wetland restored. • There was no additional progress this year. There are currently no acres of CP-23 (restored wetland) enrolled under this agreement.
		The specific measurable criteria of success will be acres of wetland restored.	
Montana - Upper Clark Fork River Basin	Water Quality	Reduce non-point source pollution, restore environmental function, and improve wildlife and habitat by improving or restoring up to 12,300 acres of riparian buffers.	• Working with a local watershed group and NRCS in 2022 and 2023, four potential projects were developed. • Also, FSA and the Montana Natural Resource Damage Program (NRDP) met with the Conservation Districts in Philipsburg and Deer Lodge to brief them on the CREP and solicit input on how to best present the program to landowners. • There were no acres enrolled under this agreement during FY 2023.
	Water Quality	Restore 1,000 acres of wetlands.	• There are currently no wetland acres enrolled under this agreement.

Agreement Name	Goal Type	Goal	Progress and Methods Used
	Habitat	Restore 5,000 acres of rare and declining habitat Sagebrush steppe.	• Working with a local watershed group and NRCS we developed four potential projects in 2022 and 2023. • Also, FSA and Montana Natural Resource Damage Program (NRDP) met with the Conservation Districts in Philipsburg and Deer Lodge to brief them on the program and solicit input on how to make it most successful. • There were no acres enrolled under this agreement during FY 2023.
	Habitat	Enroll up to 45,500 acres of grasslands to improve the hydrologic function of the uplands and provide for sustainable range management.	• The measurement for success is acres of grasslands improved. Working with a local watershed group and NRCS we developed four potential projects in 2022 and 2023. • Also, FSA and NRDP met with the Conservation Districts in Philipsburg and Deer Lodge to brief them on the program and solicit input on how to make it most successful. • There were no acres enrolled under this agreement during FY 2023.
Nebraska II-Platte Republican Resource Area	Water Quantity	Reduce the application of water for cropland irrigation in the project area by 125,000 acre-feet annually from 2004 irrigated usage levels.	• The estimated consumptive use savings for curtailing irrigation on CREP program acres for the 2023 irrigation season is 36,710 acre-feet. The implied irrigation efficiency is 0.68. Therefore, the expected reduction in application of water for 2023 is 53,985 acre-feet. This is 43 percent of the goal. • For the 2023 irrigation season, approximately 40 percent of the maximum goal acres were enrolled in the program. The progress is in line with the

Agreement Name	Goal Type	Goal	Progress and Methods Used
			level of participation. The Nebraska Legislature passed a bill in 2017 that makes re-enrollment of irrigated land under a surface water appropriation more likely, because the number of years that a surface water appropriation may be protected from cancellation for nonuse was increased from 15 to 30 years. This paved the way for 15-year contracts to be renewed for another 15 years without placing the water appropriation in jeopardy.
	Water Quantity	Increase surface and ground water retention by a target amount of 85,000 acre-feet of water annually within the project area reservoirs, groundwater tables, and streams.	• The retention of surface and groundwater is dependent and synonymous with the reduction in consumptive use. Consumptive use of irrigation water is lost to lakes, streams, and groundwater aquifers through irrigation. • The estimated retention (consumptive use) from all sources is 36,710 acre-feet for 2023, which is 43 percent of the goal. This is in line and correlated with the overall level of participation. • The Nebraska Legislature passed a bill in 2017 that makes re-enrollment of irrigated land under a surface water appropriation more likely, because the number of years that a surface water appropriation may be protected from cancellation for nonuse was increased from 15 to 30 years. This paved the way for 15-year contracts to be renewed for another 15 years without placing the water appropriation in jeopardy.
	Habitat	Provide up to 85,000 additional acres of native	• The 85,000 acres are distributed across three

Agreement Name	Goal Type	Goal	Progress and Methods Used
		grassland habitat for wildlife in the project area, increasing the populations of pheasants and other ground nesting birds by 25 percent in the area.	primary practices: Permanent Native Grasses, CP2; Permanent Wildlife Habitat, CP4D; and Rare and Declining Habitat, CP25. In addition, practice Wildlife Food Plot CP12 may be used in conjunction with any of the three primary practices. Currently all but 111.43 acres of the enrolled acres in the CREP program are in these practices. • Monitoring of wildlife populations in the Platte and Republican River Basins CREP area continues to be completed by the Nebraska Game and Parks Commission (NGPC) using standard game surveys. The primary impact on wildlife in the CREP area at this point has been the enrollment of 39,935.57 acres of formerly cropped irrigated fields into appropriate wildlife cover. The bulk of the CREP acres were enrolled in the spring of 2005 and were planted to perennial cover in the fall of 2005 and spring of 2006 and now are being reenrolled. The NGPC has graphed several species' survey data showing prevalence since 2005. The survey data provides a baseline for detecting changes in populations that can be attributed, at least in part, to CREP enrollment. Annual variations in wildlife populations are very common, and in Nebraska, are typically tied to weather conditions. Surveyed wildlife populations in the CREP area are compared to those across the state to better understand the

Agreement Name	Goal Type	Goal	Progress and Methods Used
			relative impact of CREP habitat enrollments on Nebraska wildlife populations of interest. • Pheasants improved slightly in the Platte River CREP area as did bobwhite quail likely due to dry
			• conditions during nesting and hatch. All surveyed species in the CREP area are faring as well as, or better than, the statewide average.
	Habitat	Provide up to 15,000 additional acres of conservation buffers and restored wetlands.	• There are Sign-up Incentive Payments (SIP) and Practice Incentive Payments (PIP) for applicable practices (Filter Strips CP21, Riparian Buffer CP22, Wetland Restoration CP23 and Wetland Restoration/Non-Floodplain CP23A) to encourage achieving this goal. Cover establishment is reimbursed with a 50/50 share between USDA and the State Partner to further incentivize participation. • The 15,000-acre target is further broken down to 10,000 acres for CP21 and CP22, and 5,000 acres for CP23 and CP23A. • Currently, there is only one contract for 111.43 acres of CP23 practice. It is unusual in that it has irrigated land situated partly in a wetland area. • There are currently 19.58 acres of CP21 and 1.21 acres of CP22 practices enrolled under this agreement.
	Water Quality	Seek to reduce the application of triazine products by approximately 93,000 pounds annually, when fully enrolled,	• Under the terms of the program, lands included under contract must be replanted to native grasses and, therefore,

Agreement Name	Goal Type	Goal	Progress and Methods Used
		from existing application rates in the project area.	would not be treated with herbicides. • The average amounts of application associated with triazine compounds at the inception of the program was 1.3 pounds per acre. Therefore, the amount of triazine that likely would have been applied to the contracted acres, had they remained as irrigated cropland, is approximately 52,061 lbs. • This is 56 percent of the goal, which has partially been met. This is in part due to the 40 percent participation rate for total enrolled acres.
	Water Quality	Seek to reduce leaching of nitrate compounds into project area streams and groundwater by 5,900,000 pounds annually, when fully enrolled, from the 2004 application rates.	• Under the terms of the program, lands included under contract must be replanted to native grasses and, therefore, would not be fertilized. • The average nitrogen application is 200 pounds per acre. Therefore, the amount of nitrogen that likely would have been applied to the contracted acres, had they remained as irrigated cropland, is approximately 8,009,400 lbs. • This goal has been met.
	Water Quality	Seek to reduce the application of phosphate products by approximately 2,440,000 pounds annually, when fully enrolled, from 2004 application rates in the project area.	• Under the terms of the program, lands included under contract must be replanted to native grasses and, therefore, would not be fertilized. The original goal at a full program with 100,000 acres computes to 24.4 lbs. per acre. • However, the average amount of application associated phosphate was estimated in the original program environmental review documents as 20 pounds per acre. Using the 20 lbs. per

Agreement Name	Goal Type	Goal	Progress and Methods Used
			acre value, the amount of phosphate that likely would have been applied to the contracted acres, had they remained as irrigated cropland, is approximately 800,940 lbs. • The goal was not met and, at 33 percent, is one of the furthest from being met.
	Water Quality	Assist community public water supplies (surface and groundwater) by reducing nitrogen and phosphorus levels from agricultural activities.	• Nebraska's Natural Resources Districts (NRDs) are the primary regulator of nonpoint source pollution in groundwater. NRDs develop and implement groundwater quality managements plans that describe monitoring, assessment, and thresholds triggering regulatory measures. There are many examples throughout the state of increased regulatory measures to protect and restore community public water supplies. NRDs partner with local communities, agricultural producers, and the
			• private sector to leverage resources that protect water quality. Education and outreach are offered and cost-share is available for BMPs that help producers reduce water use and fertilizer application. Nitrate levels are annually measured and monitored and trigger levels for regulations that have been implemented in several NRDs. • The Nebraska Department of Environment and Energy (NDEE) administers the Wellhead Protection Program and state Nonpoint Source Management program. Together, these programs offer planning and financial

Agreement Name	Goal Type	Goal	Progress and Methods Used
			assistance to public water systems interested in protecting and restoring their water supplies. • Recently, NDEE is funding nonpoint source planning efforts specific to community water system wellhead (source water) protection areas. Once these community-based plans are approved by EPA, implementation is eligible for federal Clean Water Act Section 319 nonpoint source funds. Eight such plans are currently approved and in development.
	Education	Provide educational assistance to project area irrigators to develop a more efficient use of applied water, nutrients, and herbicides.	• and information to assist in making effective conservation-minded management decisions. While there are special Water Quality and Quantity Management Areas where certain practices are required, many of the most effective practices being implemented by producers across the state are voluntary. Using NRD funds to leverage state and federal dollars, local boards have been able to provide cost-share incentives to producers for innovative, research-driven advances in irrigation management. • A few examples of some of the programs offered or required by NRDs are: • NRD-level cost-share programs offering technical assistance of 0-100 percent on purchase of soil moisture sensing equipment; • Required flow meters on groundwater wells over specified capacity;

Agreement Name	Goal Type	Goal	Progress and Methods Used
			• Allocation of a certain number of inches that can be pumped over a certain number of years; • Required soil sampling for water quality indicators. • A recent example of a creative program to engage producers in the adoption of new technologies is the Testing Ag Performance Solutions (TAPS) program (taps.unl.edu). This program provides opportunities for producers to virtually compete against each other as well as UNL scientists for: (1) most profitable farm, (2) highest input (water and nitrogen) use efficiency, and (3) greatest grain yield. The goal of the competition is to promote efficiency and profitability while giving a chance to learn from those who grow corn profitably. The competition is supported by UNL Extension, NRDs, non-profit organizations, and agricultural industries, among others. • The program has grown each year and has expanded to different farming scenarios. The data gathered each year is being analyzed to provide better support and recommendations for agricultural producers.
	Habitat	Monitor the aquatic communities and associated habitat parameters in project area reservoirs and rivers to determine biological relationships.	• The fisheries program with Nebraska Game & Parks Commission (NGPC) has been involved in an on- going limnological assessment at Harlan County Reservoir during the entire Nebraska CREP program timeframe. • NGPC has a consistent data base of abiotic, zooplankton

Agreement Name	Goal Type	Goal	Progress and Methods Used
			and larval fish collection results. For the Platte River basin, NGPC conducts standardized annual fish monitoring surveys on the reservoirs. In addition, NDEE facilitates water quality sampling and management statewide through EPA Clean Water Act programs. • During FY 2023, monitoring within the CREP area was conducted through 3 monitoring programs. Monitoring results are used to produce the Integrated Report that combines the 303(d) list of Impaired Waters and the 305(b) Water Quality Report every two years. The report is used for future water quality management, particularly though the National Pollution Discharge Elimination System (discharge permits for point sources) and the Section 319 Nonpoint Source Pollution grant program. Common lake impairments in the CREP area are for fish consumption, nutrients, pH, and chlorophyll, while stream impairments are predominantly for E. coli bacteria. Bacteria total maximum daily loads (TMDLs) have been established for numerous stream segments in the CREP area and more will be added. Bacteria TMDLs for 5 stream segments in the Republican River basin are in the final public comment period before being submitted to EPA. • In addition to these programs, NDEE also administers ambient stream, lake, and public beach monitoring programs.

Agreement Name	Goal Type	Goal	Progress and Methods Used
	Environ-mental	For irrigation purposes, reduce the total consumption of fossil fuels by 350,000 gallons and electricity use by 10 million kilowatt hours.	• At the inception of the program, the Nebraska Department of Energy's data indicated that approximately 55 percent of all irrigation pumps are powered by electricity, and 45 percent are powered by fossil fuels. The Nebraska Public Power District, one of the Nebraska Platte/Republican CREP partners, provided information from a 2001 Report – "Estimated Irrigation Costs" by Roger Selley, UNL. Using assumptions based upon that report, the following method has been employed each year to estimate the energy savings from the NE CREP. • The representative distribution system is a 135-acre center pivot pumping 800 gallons per minute and applying 9.5 acre-inches per acre with a lift of 100 feet at 60 percent efficiency, the annual electric usage is 45,966 kilowatt hours, and fossil fuels (diesel, propane and gasoline) average 4,600 gallons. • The formulas used are (electric consumption = acres x .55 x (46,000 kilowatt hours/135 acres)) and (fossil fuel consumption = acres x .45 x (4,600 gallons/135 acres)). Using this method, the 2023 estimated electrical energy savings would have been 7,505,104 kilowatt hours. • The estimated fossil fuels savings would have been 614,054 gallons, indicating that the fossil fuel goal was met. The electrical energy savings was about 75 percent met.

Agreement Name	Goal Type	Goal	Progress and Methods Used
New Jersey	Water Quality	Maintain and restore ecological functions of streams by reducing biological impairments.	• 13.82 stream miles were buffered through this agreement. Of all acres enrolled through the agreement, nearly half represent riparian buffer acres, indicating a strong focus on water quality and stream health.
	Water Quality	Annual reduction of 26,000 pounds of phosphorous.	• Based on the number of stream miles buffered, an estimated 1,170 pounds of phosphorus were removed from NJ watersheds last year due to buffers implemented under this agreement.
	Water Quality	Annual reduction of 7 million pounds of Total Suspended Solids (TSS).	• Based on the number of stream miles buffered, an estimated 939,652 pounds of TSS were removed from NJ watersheds last year due to buffers implemented under this agreement.
New York I-New York City	Water Quality	Reduce the amount of silt and sedimentation entering the tributaries, main streams, and reservoirs in the Catskill and Delaware watersheds of the New York City drinking water supply system.	• Progress toward achieving the goal of reducing the amount of sediments entering the New York City drinking water supply system may be estimated through modeling. The Chesapeake Assessment Scenario Tool (CAST) estimates edge of stream load reductions for TSSs using coefficients for various BMPs. The impact of riparian buffer plantings was assessed in CAST using scenarios of current acreage of riparian forest buffer plantings compared to a 2007 baseline scenario. • Riparian forest buffer plantings with exclusion fencing in the estimated 1,243 acres of CREP lands under contract as of 09/30/23 showed the potential to reduce total TSS by ~4,550 tons annually.

Agreement Name	Goal Type	Goal	Progress and Methods Used
			• In addition to CREP, the Watershed Agricultural Program (WAP) has implemented thousands of additional agricultural BMPs (including nutrient management plans, cover crops and exclusion fencing) since the NYC CREP agreement was signed in 1998 that have helped reduce the amount of sediments entering the water supply system.
	Water Quality	Reduce the amount of phosphorous and nitrogen entering the tributaries, main stems, and reservoirs in the Catskill and Delaware watersheds of the New York City drinking water supply system.	• Progress toward achieving the goal of reducing the amounts of phosphorous and nitrogen entering the New York City drinking water supply system was estimated through the modeling described above. • CAST estimated that the 1,243 acres of riparian forest buffers under contract as of 9/30/22 had the potential to reduce phosphorous by ~ 7.6 tons and nitrogen by ~30.3 tons annually.
	Water Quality	Reduce the amount of waterborne pathogens entering the tributaries, main stems, and reservoirs in the Catskill and Delaware watersheds of the New York City drinking water supply system.	• It is difficult to evaluate progress toward achieving the goal of reducing the amount of waterborne pathogens entering the New York City drinking water supply system attributable to CREP because of the size and dynamics of the watershed. • In addition to CREP, there are numerous programs and practices in the NYC watershed dedicated to controlling and monitoring pathogens, including ultra-violet treatment. Furthermore, the WAP has implemented thousands of additional agricultural best management practices since the CREP agreement was signed in

Agreement Name	Goal Type	Goal	Progress and Methods Used
			1998, including calf housing and young stock manure storage practices, which have helped reduce the amounts of pathogens from agricultural activities entering the water supply.
	Water Quality	Promote the continued health and viability of natural habitats and ecosystems in the Catskill and Delaware watersheds of the New York City drinking water supply system, including those of endangered species, such as the Bald Eagle and native cold-water fish.	• True aquatic habitat benefits for cold water species or eagle habitat will be achieved when canopy closure over CREP reaches on streams is achieved. • The Stroud Water Research Center in southeastern Pennsylvania, which has highly productive soils and a warmer, extended growing season than the NYC • Watershed, estimates reaching canopy closure in 15 years with perfect tree maintenance management.
			• WAP staff estimate that it will take at least 25 years to achieve canopy closure in the NYC watershed based on soils, climate, and management.
New York II-Syracuse	Water Quality	Reduce the risk of pathogens from animal manure applied to pasture and cropland.	• Farms voluntarily participating in the City of Syracuse Skaneateles Lake Watershed Agricultural Program (SLWAP) have a Whole Farm Plan, which contains manure spreading recommendations. The SLWAP Farm Planner obtains and reviews that data annually. The planner also generates nutrient management recommendations (manure spreading volumes and locations) for farms and crop fields.
	Water Quality	Reduce sediment deposition in Skaneateles Lake and its	• Farms voluntarily participating in the City of Syracuse SLWAP have a

Agreement Name	Goal Type	Goal	Progress and Methods Used
		tributaries attributable to erosion of croplands.	Whole Farm Plan, which helps farmers implement buffers and cover crops on their corn-for- silage ground.
	Water Quality	Reduce nutrient runoff from animal manure and fertilizer applied to adjacent cropland and pastures.	• Farms voluntarily participating in the City of Syracuse SLWAP have a Whole Farm Plan, which contains manure and fertilizer spreading recommendations. The SLWAP farm planner obtains and reviews that data annually. The planner also generates nutrient management recommendations (manure spreading and fertilizer spreading volumes and locations) for farms and crop fields.
	Water Quality	Assist the City of Syracuse in continuing to meet filtration avoidance criteria issued by the NYS Dept. of Health in order to comply with the Safe Drinking Water Act.	• Farms voluntarily participating in the City of Syracuse SLWAP have a Whole Farm Plan, which contains BMPs that the farmer and the SLWAP have agreed to plan and implement. The City of Syracuse provides funding for the SLWAP. The SLWAP staff leverage those funds to secure additional grant funding to implement BMPs. • The SLWAP Farm Planner reviews 100 percent of the BMPs on 25 percent of the farms annually. Any BMPs that need repair, replacement, etc. are re- planned with the farmer, then City of Syracuse funds
			are again leveraged to secure additional grant funds to repair and/or replace the BMP.
	Habitat	Establish, preserve, or enhance wildlife habitat leading to an increase in populations and diversity of	• The SLWAP staff work with farmers in the watershed to plan and implement BMPs related to wildlife and habitat.

Agreement Name	Goal Type	Goal	Progress and Methods Used
		birds, mammals, and aquatic organisms.	Some farmers in the past have participated in USDA's delayed mowing program. • Other farms have installed 25 acres of buffer strips around their forest lands and/or hedgerows that increase habitat diversity and provide permanent vegetative cover.
New York III-Statewide	Water Quality	Reduce phosphorous from 145,284 lbs. per year to 72,642 lbs. per year.	• Phosphorus reductions for CREP practices contracted in 2023 are estimated by using the Chesapeake Assessment Scenario Tool 17 (CAST 17, Chesapeake Bay Program). • For FY 2023, phosphorus reductions were estimated at 6,412 lbs. (short of the goal).
	Water Quality	Reduce nitrogen from 77,316 lbs. per year to 38,688 lbs. per year.	• Nitrogen reductions for CREP practices were estimated by using the Chesapeake Assessment Scenario Tool 17 (CAST 17, Chesapeake Bay Program). • For FY 2023, nitrogen reductions were estimated at 43,220 lbs. (exceeding the goal).
	Water Quality	Reduce sedimentation from 175,316 tons per year to 70,216 tons per year.	• Sediment reductions for CREP practices were estimated by using the Chesapeake Assessment Scenario Tool 19 (CAST 17, Chesapeake Bay Program). Values for sediment were calculated using the average of reduction coefficients within CAST 17 for pastureland and a range of values available for cropland. • For FY 2023, sediment reductions were estimated at 562 tons (short of the goal).
	Water Quality	Establish riparian buffers adjacent to 4,598 stream miles.	• Progress towards establishing riparian buffers adjacent to 4,598 stream miles is not measured or monitored. Miles of riparian buffer is not the

Agreement Name	Goal Type	Goal	Progress and Methods Used
			reportable unit for that practice standard (NRCS). Cumulative progress towards this goal is estimated at 526 miles in total and 4.95 miles from Riparian • Buffers implemented in FY 2023 as complimentary practices.
	Water Quality	Establish riparian buffers adjacent to 473,457 acres of surface waters.	• Progress towards establishing riparian buffers adjacent to 473,457 acres of surface waters is not measured or monitored. Acres of surface waters adjacent to riparian buffers is not the reportable unit for that practice (NRCS).
	Water Quality	Establish conservation cover on areas that serve EPA-approved wellhead zones.	• Progress towards establishing conservation cover on EPA-approved wellhead zones is minimal.
North Carolina	Water Quality	Restore and enhance riparian habitat corridors next to streams, drainage ditches, estuaries, wetlands, and other water courses by enrolling up to 85,000 acres of riparian forested buffers, grass filter strips and other riparian tree plantings.	• A total of 30,000 acres is currently in the North Carolina CREP. Though there are only 3,165 acres currently under contract with FSA, this number includes expired CRP contracts that also had a conservation easement option. Even though the CRP contract expired, the conservation easements remain for the acquired term (30-year or permanent). In six more years, the first group of 30- year easements expire. • With new enrollment of 34.6 acres, 37 tons per year of soil is saved; nitrogen reduction is 1,755.84 pounds per year and phosphorus reduction is 3.07 pounds per year. (Source: North Carolina Agriculture Cost Share Program BMP Results Report.) • Three conservation easements were completed (closed) during FY 2023. Two are 30-

Agreement Name	Goal Type	Goal	Progress and Methods Used
			year easements and one is a permanent easement for a total of 42.031 acres. Soil loss and nutrient reductions for these conservation easements would have already been reported in previous years when riparian buffers were established. The closing of the conservation easements takes several years to complete, but these easements help to lengthen the timeframe that riparian buffer zones will be in place.
	Habitat	Restore up to 15,000 acres of non-riparian wetlands either associated with drainage ditches or adjacent to primary fishery nursery areas to address impacts associated with drainage.	• 2,820 acres have been enrolled towards meeting this goal. • There have been no new enrollments using CP-23 in the past 2 fiscal years. With no new enrollments, there are no soil loss or nutrient reductions to report.
	Water Quality	Help farmers comply with the Neuse and Tar- Pamlico Nitrogen Reduction Rules and potential regulations or goals in other watersheds.	• CREP continues to provide an opportunity for farmers to establish riparian buffers that help to slow the water down and allow for sediment and nutrients to settle before the water enters streams. Riparian buffer zones used in conjunction with changes in management practices on the farm have allowed agriculture to exceed the required 30 percent reduction of nitrogen in several river basins. Agriculture has been able to report a reduction in nitrogen annual loads of 55 percent in the Neuse River basin (compared to a baseline of agriculture fields in 1991-1995) and a 57 percent reduction in the Tar-Pamlico River Basin (compared to a 1991 baseline). (Source: 2023 Annual Progress Report for

Agreement Name	Goal Type	Goal	Progress and Methods Used
			the Neuse and Tar-Pamlico Agricultural Rule.) • Traditionally, CREP projects have a large presence in eastern NC due to the Neuse and Tar-Pamlico River basins. As other river basins and local watersheds around the state (within the CREP area) develop new nutrient reduction rules, CREP provides the first step in helping farmers reach their goals. Promotional programs are being developed to better assist and provide information to farmers and natural resource professionals in these areas, including High Rock Lake Watershed and the Jordan Lake Watershed.
			• There are no measurable goals to report for other watersheds at this time. CREP staff are looking to improve reporting methods of miles of streams protected in individual watersheds.
North Dakota Riparian Project	Habitat	Enroll 20,000 acres of cropland in the Conservation Reserve Program (CRP) consisting of filter strips, riparian buffers, and pollinator habitat to improve and maintain water quality and wildlife habitat, including habitat and forage for pollinators.	• 1,051.2 acres have been enrolled to date, well short of the desired goal. These acres are providing habitat for all species.
	Habitat	Enroll 40,000 acres of land into the North Dakota Game and Fish Department's Private Land Open to Sportsmen (PLOTS) program.	• 2,444 acres of CREP and non-CREP have been enrolled towards this goal to date.
	Water Quality	Improve water quality in the project area by enrolling 1,500 acres annually reducing the amount of nitrogen, phosphorus and sediment	• For fiscal year 2023, annual reduction of nitrogen, phosphorous, and sediment are as follows:

Agreement Name	Goal Type	Goal	Progress and Methods Used
		entering rivers, streams, and other waterbodies by 75,000 pounds of nitrogen per year; 37,500 pounds of phosphorus per year; and 4,750 tons of sediment per year.	- 140,884 pounds nitrogen - 14,087.6 pounds phosphorous - 1,759 tons of sediment
Ohio I-Lake Erie	Water Quality	Install 67,000 acres in conservation practices on 10 percent of the total riparian acres in the basin.	• Progress is measured by the number of practices that are installed annually. • During FY 2023, 1,123 acres of new practices were enrolled and over 1,600 acres were re-enrolled. • The new re-enrollment incentive is expected to help sustain existing acres while encouraging enrollment of new acres.
	Habitat	Increase the population of migratory birds and waterfowl, grassland birds, and other wildlife by 50 percent when compared to similar land managed for agricultural production.	• An inventory of migratory birds will be conducted every other year in the Lake Erie CREP project area. The inventory will allow for a comparison of populations of migratory birds found on the properties enrolled within the Lake Erie CREP with those of similar acres managed for agricultural production. • No inventory was completed for 2023, but a report will be provided for 2024.
	Water Quality	Maintain enrollment to sustain total suspended solids levels at the rate of 135 mg/L as an annual flow-weighted mean concentration.	• Progress is measured by data collected by Heidelberg College through water quality sampling. Final flow values from USGS generally do not become available until at least a year after the end of the water year of interest. Since these data sets are needed to calculate loads, calculations for the most recent year are based on provisional data and are subject to revision. • For FY 2022, total suspended solids demonstrated an overall

Agreement Name	Goal Type	Goal	Progress and Methods Used
			increase, reported at a rate of 146 mg/L as an annual flow-weighted mean concentration. An increase in TSSs occurred across all monitored watersheds as compared to previous years. This increase could be due to many factors, which is why it is best to look at an average of several years. The average TSSs for this project area during the past 10 years is just below the goal of 135 mg/L.
Ohio III-Scioto River Watershed	Water Quality	Create 70,000 acres of permanent vegetative cover adjacent to rivers and streams when full enrollment is reached.	• Progress is measured by the acreage in practices installed annually. For FY 2023, 2,075 acres of new practices were enrolled and 6,566 acres were re- enrolled. • With changes made to the Scioto River Watershed (SRW) CREP, including both federal and non-federal incentives, we expect enrollment to continue to increase in FY 2024.
	Habitat	Increase the population of migratory birds and waterfowl, grassland birds, and other wildlife on land enrolled in CREP by 50 percent, when compared to similar land managed for agriculture production.	• In FY 2023, the Ohio Department of Natural Resources (ODNR) compared lands enrolled in the SRW CREP to nearby croplands and measured bird abundance on grasslands and in crop fields. During each survey, staff recorded all birds seen and heard within 100 meters of the survey location. Species, sex, and distance to each bird was recorded.
			• In total, 25 species of grassland birds were detected across 44 point-count surveys. Of the 25 species detected, 22 species were found in CREP fields and 15 were found in croplands. A total of 332 individual birds were

Agreement Name	Goal Type	Goal	Progress and Methods Used
			recorded, with 260 individual birds detected in CREP fields and 72 individuals detected in croplands. • Average grassland bird species richness was 146 percent greater in CREP fields than in nearby croplands. Average grassland bird diversity was 1.28 times greater in CREP fields. The total estimated density of grassland birds was nearly 2.5 times greater in CREP fields. • In short, the SRW CREP is exceeding the goal of increasing wildlife populations by 50 percent.
	Water Quality	Maintain enrollment to sustain total suspended solids levels at the rate of 85 mg/L as an annual flow-weighted mean concentration.	• Progress is measured by data collected by Heidelberg College through water quality sampling. Final flow values from USGS generally do not become available until at least a year after the end of the water year of interest. Since these data sets are necessary to calculate loads, calculations for the most recent year are based on provisional data and are subject to revision. • The FY 2022 results show an overall increase of TSSs reported at a rate of 91 mg/L as an annual flow- weighted mean concentration. TSSs increased across all monitored watersheds in FY 2022 when compared to previous years. This increase could be due to many factors, which is why an average is used over several years to determine trends in water quality. • The average TSSs in recent years is 79 mg/L, which is below the goal of 85 mg/L.

Agreement Name	Goal Type	Goal	Progress and Methods Used
Oregon	Water Quality	Restore 100 percent of the area enrolled for the riparian forest buffer practice to a properly functioning condition in terms of distribution and growth of woody plant species, filtration of nutrients and sediment from agricultural runoff, and shade and stabilization of stream banks under normal non-flood conditions.	• The Oregon Watershed Enhancement Board (OWEB) has been working with CREP technicians to track contract performance and inform management of CREP buffers. This approach includes completion of a CREP monitoring checklist and repeating the Stream Visual Assessment Protocol (SVAP) on a subset of active contracts. The technicians will report their findings in early 2024, and OWEB will continue to incorporate lessons learned to continually refine this monitoring approach.
	Water Quality	Provide a way for farmers and ranchers to voluntarily meet water quality requirements established under federal law and under Oregon's water quality act.	• CREP technicians continue to perform outreach to landowners to enroll/re-enroll in CREP. They meet on-site with landowners and complete contracts for re-enrollments, new enrollments, and conservation plans. In addition to status reviews to verify producers are implementing the practices they're getting paid for, CREP technicians employ adaptive management when needed to get plantings up to desired survival rate, as well as performing SVAPs.
	Water Quality and Habitat	Attain enrollments for the following targets for a maximum of 100,000 acres enrolled in Oregon under CREP: Coastal Basins - 1,250 acres of riparian forest buffer; 1,000 acres of restored wetlands = 2,250 acres Columbia Basin - 8,000 acres of riparian forest buffers; 1,000 acres of restored wetlands = 9,000 total acres	• Based on our enrollment numbers, Oregon has achieved the goals set forth in the various regions outlined above for the state. • There are 22,673.75 total acres of riparian buffer (CP22) and 77.44 acres of restored wetlands (CP23) enrolled across this agreement.

Agreement Name	Goal Type	Goal	Progress and Methods Used
		Interior drainages - 3,500 acres of riparian forest buffers, 1,000 acres of restored wetlands = 4,500 total acre.	
	Habitat	This is the primary goal. A number of fish species native to Oregon have either been listed or proposed for listing as threatened or endangered species under the Federal Endangered Species Act. This Agreement seeks to help alleviate some of these problems.	• As a result of the nearly 48,000 acres that are enrolled in CREP, the water temperatures and shade provided by riparian buffer practices, in addition to filtering agricultural runoff, have provided improved habitat for a number of fish species located in the state.
Pennsyl-vania I-Chesapeake Bay Watershed	Water Quality and Habitat	Permit the Commonwealth's producers to voluntarily restore wetlands, riparian areas, and grasslands by enrolling up to 219,746 acres of farmland into the Chesapeake Bay CREP, through financial and technical assistance.	• Pennsylvania currently has 55,769.7 acres of wetlands, riparian areas, and grasslands enrolled in the Chesapeake Bay CREP. • The sum of acreage for wetlands, riparian areas, and grasslands used conservation practices CP1, CP2, CP8A, CP9, CP10, CP15A, CP21, CP22, CP23, CP29, • CP30, and CP33 based on the Chesapeake Assessment Scenario Tool (CAST) "category." • Conservation practices CP3A, CP4D and CP12 were excluded from the calculation.
	Water Quality	Reduce erosion on cropland in the Chesapeake Bay watershed by an estimated 3.5 million tons using April 22, 2000, as the beginning of the base period for measuring erosion reduction levels.	• Pennsylvania's Chesapeake Bay CREP is responsible for preventing an estimated 32,123 tons/yr. of sediment from entering the Chesapeake Bay Watershed during this program year. • Estimated calculations used loading rates and reduction coefficients based on the Chesapeake Bay Watershed Model.
	Water Quality	Prevent 72,500 tons of sediment and 4.4 million pounds of nitrogen and	• Estimated pollution prevented from entering the Chesapeake

Agreement Name	Goal Type	Goal	Progress and Methods Used
		phosphorus from reaching the Chesapeake Bay using April 22, 2000, as the beginning of the base period for measuring erosion reduction levels.	Bay Watershed during this program year: • 32,123 tons/yr. of sediment • 1,581,397 lbs./yr. of nitrogen • 84,198 lbs./yr. of phosphorous • Estimated calculations use loading rates and reduction coefficients based on the Chesapeake Bay Watershed Model.
	Water Quality and Habitat	Restore and enhance riparian habitat corridors next to streams, estuaries, wetlands, and other watercourses by seeking to enroll at least 31,746 acres of buffers, grass filter strips, and wetlands.	• Pennsylvania currently has 13,337.9 acres of buffers, grass filter strips, and wetlands enrolled in the Chesapeake Bay CREP.
			• The sum of acreage for buffers, grass filter strips, and wetlands used conservation practices CP8A, CP9, CP15A, CP21, CP22, CP23, CP29, CP30, and CP33 based on the CAST "category." Conservation practices CP1, CP2, CP3A, CP4D, CP10 and CP12 were excluded from the calculation.
	Water Quality and Habitat	Restore and enhance grassland habitats for declining grassland-dependent wildlife and improve water quality by seeking to enroll 188,000 acres of Highly Erodible Land (HEL) in conservation cover plantings.	• Pennsylvania currently has 58,006.4 acres of HEL in conservation cover plantings through the Chesapeake Bay CREP.
	Water Quality and Habitat	Improve water quality and restore damaged riparian areas of the Susquehanna and Potomac Watersheds to facilitate the health of fish, game, and other wildlife populations.	• Under the Chesapeake Bay CREP, Pennsylvania currently has 13,337.9 acres of buffers, grass filter strips, and wetlands and 58,006.4 acres of HEL in conservation cover plantings.
	Water Quality	Permit producers to voluntarily restore and protect	• Pennsylvania currently has 14,499.80 acres of cropland

Agreement Name	Goal Type	Goal	Progress and Methods Used
Pennsylvania II- Ohio River Basin	and Habitat	wetlands, highly erodible land, and riparian areas by enrolling up to 40,000 acres of cropland.	enrolled in the Ohio River Basin CREP.
	Water Quality	Reduce erosion on cropland by an estimated 10,154 tons/year from 03/04/2004 levels.	• Pennsylvania's CREP program is responsible for preventing an estimated 4,996.13 tons of sediment from entering the Chesapeake Bay Watershed during this program year. This estimate is based on FSA data and uses loading rates and reduction coefficients based on a Chesapeake Bay Watershed model.
	Water Quality	Reduce 10,154 tons of sediment, 738,500 pounds of nitrogen, and 16,000 pounds of phosphorus per year from entering the Ohio River and Gulf of Mexico from 03/04/2004 levels.	• The estimated pollution prevented from entering the Ohio River and Gulf of Mexico during this program year are: • 4,996.13 tons/yr. of sediment • 271,018 lbs./yr. of nitrogen • 8,005 lbs./yr. of phosphorous • These estimates are based on FSA data and use loading rates and reduction coefficients based on a Chesapeake Bay Watershed model.
	Water Quality and Habitat	Restore and enhance riparian habitat corridors next to streams, estuaries, wetlands, and other watercourses by seeking to enroll at least 6,000 acres of buffers, grass filter strips, and wetlands.	• Pennsylvania currently has 3,828.4 acres of buffers, grass filter strips, and wetlands enrolled in the Ohio River Basin CREP.
	Water Quality and Habitat	Restore the cover of up to 34,000 acres of Highly Erodible Land (HEL) to protect water quality and create wildlife habitat, particularly grassland habitat for song- and ground-nesting birds.	• Pennsylvania currently has 10,671.4 acres of highly erodible land (HEL) enrolled in the Ohio River Basin CREP.

Agreement Name	Goal Type	Goal	Progress and Methods Used
Pennsylvania III - Delaware River Basin	Water Quality and Habitat	Permit Pennsylvania landowners and operators to voluntarily restore and protect wetlands, Highly Erodible Land (HEL), and riparian areas by enrolling up to 20,000 acres of farmland in the Delaware River Basin CREP.	• Pennsylvania currently has 90.3 acres enrolled in the Delaware River Basin CREP.
	Water Quality	Reduce erosion in the Delaware River, and ultimately the Delaware Bay, by an estimated 557 tons per year when fully enrolled.	• The Delaware River Basin CREP is responsible for preventing an estimated 13 tons of sediment from entering the Delaware River, and ultimately the Delaware Bay, during this program year. • This calculation uses loading rates and reduction coefficients based on the Chesapeake Bay Watershed Model.
	Water Quality	Prevent 557 tons of sediment, 349,500 pounds of nitrogen, and 12,353 pounds of phosphorus per year from entering the Delaware River and Bay when fully enrolled.	• Estimated pollution prevented from entering the Delaware River Basin during this program year: • 13 tons/yr. of sediment • 1,098 lbs./yr. of nitrogen • 32 lbs./yr. of phosphorous • These calculations use loading rates and reduction • coefficients based on the Chesapeake Bay Watershed Model.
	Water Quality and Habitat	Restore up to 4,000 acres of riparian buffers and wetlands; this will reduce flooding, improve in-stream water quality, stabilize stream banks and flood plains, reduce water temperature, increase time to runoff, facilitate groundwater recharge,	• Pennsylvania currently has 3.5 acres of buffers, grass • filter strips, and wetlands enrolled in the Delaware River Basin CREP. • The sum of the acreage for buffers, grass filter strips, and wetlands used conservation practices CP8A, CP9, CP15A, CP21, CP22, CP23, CP29, CP30, and
		and provide a host of other benefits to humans and wildlife.	• CP33 based on the Chesapeake Assessment Scenario Tool (CAST) "category." Conservation practices CP1,

Agreement Name	Goal Type	Goal	Progress and Methods Used
			• CP2, CP3A, CP4D, CP10 and CP12 were excluded from the calculation.
	Water Quality and Habitat	Restore up to 16,000 acres of HEL to protect water quality and create wildlife habitat, particularly grassland habitat for song- and ground-nesting birds, by planting native warm-season grasses and/or cool-season grasses, creating field borders, and protecting intact habitats.	• Pennsylvania currently has 86.9 acres of HEL enrolled in the Delaware River Basin CREP.
South Dakota- Big Sioux River Watershed	Habitat	Create 25,000 acres of permanent vegetative cover by restoring wetland and grasslands to store water and serve as breeding habitat for migratory and resident wetland- and grassland- dependent species.	• Over 784 acres of permanent vegetation has been created providing prime nesting and breeding habitat for migratory and resident wildlife.
	Water Quality	Establish 1,500 acres of permanent vegetative cover adjacent to rivers and streams.	• Over 536 acres of permanent vegetative cover have been established adjacent to rivers and streams in the Big Sioux River Watershed. • This is measured and monitored by performing a spatial analysis in Geographic Information System software of the acres of land enrolled in the Big • Sioux River watershed CREP that are adjacent to a river or stream.
	Water Quality	Provide and maintain a reduction of sediment, phosphorus, and nitrogen runoff from agricultural land previously used for row crops as follows: by 90 percent from 20,000 tons/year to 2,000 tons/year; phosphorous by 75 percent from 33,500 lbs./year to 8,375 lbs./year; and nitrogen by 85 percent from 112,500 lbs./year to 16,875 lbs./year.	• Sediment has been reduced by 95 percent from 1,058 tons/year to 53 tons/year. • Phosphorus has been reduced by 92.8 percent from 1,435 lbs./year to 103 lbs./year. • Nitrogen has been reduced by 93.4 percent from 4,915 lbs./year to 324 lbs./year on the 784 acres of cropland enrolled. • This is measured by running each piece of land enrolled in

Agreement Name	Goal Type	Goal	Progress and Methods Used
			CREP through the Spreadsheet Tool for • Estimating Pollutant Loads (STEPL) model to estimate reductions. These lands are maintained in perennial vegetative cover and comply with the CRP contract.
	Wildlife Produc- tion	Produce an additional 71,250 pheasants annually.	• According to the 2006 report by Nielson et al., • "Estimating Response of Ring-necked Pheasant (Phasianus colchicus) to the Conservation Reserve Program," 25,000 acres of CRP is estimated to produce 71,250 pheasant annually. • Based on the number of acres enrolled in CREP, it is estimated that over 2,234 additional pheasants are produced annually.
	Wildlife Product- ion	Contribute an additional 15,000 ducks to the fall flight annually.	• According to the 2007 report by Reynolds and • Shaffer, "Waterfowl Response to the Conservation Reserve Program and Swamp buster Provision in the Prairie Pothole Region, 1992-2004," 25,000 acres in CRP are estimated to produce 15,000 ducks annually. • Based on the number of acres enrolled in CREP, it is estimated that over 1,300 additional ducks are produced annually.
	Public Recrea- tion	Open 250 40-acre blocks of private land to public hunting and fishing access.	• This is measured and monitored by performing a spatial analysis in Geographic Information System software of the acres of land enrolled in the Big Sioux River watershed CREP. • There were 11 blocks of CREP open to public hunting and fishing access during FY 2023.

Agreement Name	Goal Type	Goal	Progress and Methods Used
South Dakota-James River	Water Quality	Establish 15,000 acres of permanent vegetative cover adjacent to rivers and streams.	• Over 37,000 acres of permanent vegetative cover have been established adjacent to rivers and streams in the James River Watershed. • This is measured and monitored by performing a spatial analysis in Geographic Information System software of the acres of land enrolled in the James River watershed CREP that are adjacent to a river or stream.
	Water Quality	Provide and maintain a reduction of sediment, phosphorous, and nitrogen runoff from agricultural land previously used for row crops, if not enrolled through this CREP, as follows: by 90 percent for sediment from 80,000 tons/year to 8,000 tons/year; by 75 percent for phosphorus from 134,000 lbs./year to 33,500 lbs./year; and by 85 percent for nitrogen from 450,000 lbs./year to 67,500 lbs/year.	• Sediment has been reduced by 96.7 percent from 68,402 tons/year to 2,257 tons/year; phosphorus has been reduced by 94.3 percent from 104,193 lbs./year to 5,939 lbs./year; and nitrogen has been reduced by 86.2 percent from 348,375 lbs./year to 48,076 lbs./year on the 79,537 acres of cropland enrolled. • This is measured by running each piece of land enrolled in CREP through the Spreadsheet Tool for Estimating Pollutant Loads (STEPL) model to estimate reductions. These lands are maintained in perennial vegetative cover and comply with the CRP contract.
	Wildlife Production	Produce an additional 285,000 pheasants annually.	• According to the 2006 report by Nielson et al., who estimated response of the ring-necked pheasant (Phasianus colchicus) to CRP, 100,000 acres of CRP are estimated to produce 285,000 pheasants annually. • Based on the number of acres enrolled in CREP, it is estimated that over 226,680 additional pheasants are produced annually.

Agreement Name	Goal Type	Goal	Progress and Methods Used
	Wildlife Produc- tion	Contribute an additional 60,000 ducks to the fall flight annually.	• According to the 2007 report by Reynolds and • Shaffer, "Waterfowl Response to the Conservation Reserve Program and Swampbuster Provision in the Prairie Pothole Region, 1992-2004," 100,000 acres of CRP are estimated to produce 60,000 ducks annually. • Based on the number of acres enrolled in CREP, it is estimated that over 47,700 additional ducks are produced annually.
	Habitat	Create 85,000 acres of permanent vegetative cover by restoring wetlands and grasslands to store water and serve as breeding habitat for migratory and resident wetland- and grassland-dependent species including bobolink, upland sandpiper, chestnut collared longspur, western meadowlark,	• This is being measured and monitored by the number of acres enrolled in the James River Watershed CREP, all of which help meet this goal. • Over 79,537 acres of permanent vegetation have been created, providing prime nesting and breeding habitat for migratory and resident wildlife.
		grasshopper sparrow and savannah sparrow, dickcissel, and sedge wren.	• The South Dakota Department of Game, Fish and Parks completed a research study from January 1, 2013, to December 31, 2015, in which it contracted with two PhD students attending South Dakota State University. The objectives of the study were to assess the benefits of CREP to grassland nesting birds. Three years of field work were completed for this study with final reports completed in the spring of 2017 and included in the FY 2017 CREP report. In short, the researchers observed a rapid response in game and non-game grassland bird populations using lands enrolled in the CREP.

Agreement Name	Goal Type	Goal	Progress and Methods Used
			• They conducted observations on 693 breeding bird transect surveys and documented 3,081 individuals consisting of 49 species. Seven of eight CREP focal bird species of conservation concern were documented (excepting chestnut collared longspur). They also observed several other non-game species, as well as ring-necked pheasant, sharp-tailed grouse, and 9 waterfowl species. Nest searching of 144 fields yielded 731 nests of 28 different species, including the same 7 CREP focal species of conservation concern.
	Habitat	Open 1,000 40-acre blocks, as defined for the purposes of this CREP in Section IV. K, of private land to public hunting and fishing access.	• This is measured and monitored by performing a spatial analysis in Geographic Information System software of the acres of land enrolled in the James River watershed CREP. • There were 1,002 blocks of CREP open to public hunting and fishing access during FY 2023.
Vermont	Water Quality	Supplement existing efforts to achieve phosphorus reductions attributable to non-point sources (NPS) described in the Lake Champlain Basin Program (LCBP). The LCBP identifies a NPS phosphorus reduction target of 48.3 tons per year.	• The purpose of the program is to supplement all other ongoing efforts throughout the state to meet their respective TMDL goals. Vermont has conservation assistance programs available through state, federal and private organizations; CREP is just one piece of the portfolio of programs available to improve and protect water quality. Based on the total acres enrolled through FY 2023in the Lake Champlain Basin, the total phosphorous reduction attributed to CREP buffers is currently 8.02

Agreement Name	Goal Type	Goal	Progress and Methods Used
			metric tons per year or 3.8 percent of the required reduction for Vermont across all sectors and 5.8 percent of the required reduction for the agricultural sector.
	Water Quality	Assist in achieving nitrogen reductions attributable to NPS required by the Long Island Sound TMDL. The TMDL identifies a nitrogen reduction target of 1,173 tons per year for Massachusetts, New Hampshire, and Vermont. The TMDL provides that point sources and NPS must show a 25 and 10 percent reduction, respectively.	• In the Connecticut River watershed, where nitrates are the primary nutrient of concern, CREP is responsible for an estimated annual loading reduction of 6,318.20 lbs. per year. • The state does not currently have a non-point source nitrogen reduction tracking system for the Long Island Sound as the state is grouped with other states in the Upper Connecticut River and only estimated to contribute 4 percent of the overall loading to the watershed.
	Water Quality	Supplement existing efforts to achieve phosphorus reductions in the Lake Memphremagog Basin by 10 percent.	• In the lake Memphremagog Basin (St. Francois River), phosphorous reduction attributed to CREP buffers is currently 0.174 metric tons per year, or 1.4 percent of the required reduction for Vermont across all sectors.
	Habitat	Provide secondary benefits to wildlife and aquatic habitat.	• Habitat is not measured in quantifiable number like water quality; however, the Vermont CREP program continues to partner with US Fish and Wildlife, The Orianne Society, Trout Unlimited, The Nature Conservancy, and others to promote shared wildlife conservation goals. • The program continues to be the largest riparian forest buffer and habitat conservation program in terms of acres protected annually available to landowners in Vermont. This program continues to be a

Agreement Name	Goal Type	Goal	Progress and Methods Used
			vitally important tool to meet water quality and wildlife goals.
Virginia I-Southern Rivers	Water Quality	Enrollment of 15,500 acres of riparian forest buffers and grass filter strips.	• To date, approximately 16,016 acres of buffer (103 percent of the goal) have been restored through this CREP agreement, protecting 922 miles of stream buffered.
	Water Quality	Restore 1,500 acres of cropped wetlands.	• Thus far, less than 1 percent of Southern Rivers CREP signups have been for CP-23 Wetland Restoration on cropland with a total of 16 completed contracts. • Historically, CREP in Virginia has been much more popular on marginal pastureland in combination with stream exclusion than any practices related to the conversion of active cropland to other types of cover, including wetlands conversion.
	Water Quality	Enrollment of 3,000 acres of permanent conservation easements on land enrolled in CREP.	• Most CREP easements were enrolled between 2004 and 2011 with over 1,745 acres STATEWIDE enrolled for the Chesapeake Bay and Southern Rivers.
	Water Quality	Provide a means to reduce nutrients and sediment loading due to agricultural land use in the targeted watersheds.	• Beginning July 1, 2022, Virginia's two-year budget made $6 million available statewide to fund CREP. State cost share rates for CREP practices remain at 50%, which in combination with federal cost-share brings participant reimbursement to approximately 100% of approved costs.
	Water Quality	The VA Southern Rivers' CREP Proposal identifies the following nutrient and sediment reduction objectives based on full implementation of the program:	• Sediment reductions for CREP practices installed in FY 2023 are approximately 1,321 tons. Sediment reductions for CREP practices installed since July 1,

Agreement Name	Goal Type	Goal	Progress and Methods Used
		- Reduce sediment by 29,221 tons/year - Reduce nitrogen by 196,893 pounds/year - Reduce phosphorus by 47,472 pounds/year	2000, are approximately 316,620 tons. • Nitrogen reductions for CREP practices installed in FY 2023 are approximately 7,186 pounds. Nitrogen reductions for CREP practices installed since July 1, 2000, are over 1.7 million pounds. • Phosphorus reductions for CREP practices installed in FY 2023 are approximately 1,326 pounds. • Phosphorus reductions for CREP practices installed since July 1, 2000, are approximately 332,465 pounds.
	Water Quality	Provide an opportunity for farmers and ranchers in VA to voluntarily establish riparian and wetland areas with financial and technical assistance.	• CREP funding and technical assistance is available in all hydrologic unity in the Southern Rivers, with the exception of a few hydrologic units in a single county.
	Habitat	Provide substantial wildlife habitat enhancement for the preservation of natural diversity of the biological resources of the Commonwealth including threatened and endangered species associated with riparian and wetland habitats.	• Approximately 16,016 acres of buffer have been restored and 922 miles of stream buffered, creating and enhancing riparian habitat for many species, including threatened and endangered species.
Virginia II-Chesapeake Bay	Water Quality	Enrollment of 22,000 acres of riparian forest buffers and grass filter strips.	• To date, approximately 18,652 acres of buffer (84.8 percent of the goal) have been restored under the Chesapeake Bay CREP agreement, protecting 980 miles of stream buffered.
	Water Quality	Restore 3,000 acres of cropped wetlands.	• Thus far, the CP-23 Wetland Restoration practice has not been popular in Virginia. Only 10 CP-23 BMPs have been implemented in the history of this CREP agreement, less than half a percent of total CREP signups.

Agreement Name	Goal Type	Goal	Progress and Methods Used
			• Historically, CREP in Virginia has been much more popular on marginal pastureland in combination with stream exclusion than any practices related to the conversion of active cropland to other types of cover, including wetland conversion.
	Water Quality	Enrollment of 6,000 acres of CREP easements.	• Most CREP easements were enrolled between 2004 and 2011 with over 1,745 acres STATEWIDE enrolled (Chesapeake Bay and Southern Rivers). • In the Chesapeake Bay, approximately $715,610 in state funding has gone towards easements.
	Water Quality	Provide a means to reduce nutrient and sediment loading due to agricultural land use in the targeted watersheds and assist VA in meeting the overall	• Beginning July 1, 2022, Virginia's two-year budget made $6 million available statewide to fund CREP. State cost share rates for CREP practices remain at
		Chesapeake Bay Program goal of recusing controllable nutrient and sediment loading to the basin by 40%.	• 50%, which in combination with federal cost-share brings participant reimbursement to approximately 100% of approved costs.
	Water Quality	The VA Chesapeake Bay CREP Proposal identifies the following nutrient and sediment reduction objectives based on full implementation of the program: - Reduce sediment by 33,188 tons/year - Reduce nitrogen by 516,873 pounds/year - Reduce phosphorus by 66,953 pounds/year	• Sediment reductions for CREP practices installed in FY 2023 are approximately 29 tons. Sediment reductions for CREP practices installed since July 1, 2000, are approximately 172,362 tons. • Nitrogen reductions for CREP practices installed in FY 2023 are approximately 157 pounds. Nitrogen reductions for CREP practices installed since July 1, 2000, are over 936,210 pounds. • Phosphorus reductions for CREP practices installed in FY 2023 are approximately 24 pounds. Phosphorus

Agreement Name	Goal Type	Goal	Progress and Methods Used
			reductions for CREP practices installed since July 1, 2000, are approximately 175,163 pounds.
	Water Quality	Provide an opportunity for farmers and ranchers in VA to voluntarily establish riparian and wetland areas with financial and technical assistance.	• CREP funding and technical assistance is available in all hydrologic unity in the Chesapeake Bay watershed portion of Virginia.
	Habitat	Provide substantial wildlife habitat enhancement for the preservation of natural diversity of the biological resources of the Commonwealth including threatened and endangered species associated with riparian and wetland habitats.	• Approximately 18,651 acres of buffer have been restored and 980 miles of stream buffered, creating and enhancing riparian habitat for many species, including threatened and endangered species.
Washington	Habitat	Restoration of 100 percent of the area enrolled for the riparian forest practice to a properly functioning condition in terms of distribution and growth of woody plant species.	• A report released by the Washington State Conservation Commission states that 23,279 trees and shrubs were planted during FY 2023 and estimates that 3,144,064 trees and shrubs have been planted on current and past CREP contracts under this agreement. • Annual monitoring evaluates growth rate, diversity, and percent invasive species of conifer, deciduous, and shrub species on CREP sites in the project area.
			• Annual reports for the past 5 years have shown that most sites have growth rates appropriate for their location, or better, indicating an appropriate species mix for optimal riparian forest function in these sites. • Monitoring has shown that species diversity stabilizes over time as buffers mature.
	Water Quality	Reduction of sediment and nutrient pollution from	• A recent literature review by the Washington Department

Agreement Name	Goal Type	Goal	Progress and Methods Used
		agricultural lands adjacent to riparian buffers by more than 50 percent.	of Ecology noted that most sediment trapping in a buffer tends to occur in the first few meters (Lee et al. 2003, Zhang et al., 2009; Gharabaghi et al., 2006; Dosskey et al., 2002). In particular, Yuan et al., (2009) found that a 5-meter buffer can trap about 80 percent of incoming sediment. • Buffers in the Washington CREP program average 142 ft (433.28 m) width, greatly exceeding the 80 percent reduction afforded by 5 meters and the 50 percent reduction goal.
	Water Quality	Establishment of adequate vegetation on enrolled riparian areas to stabilize 90 percent of stream banks under normal (non-flood) water conditions.	• Annual monitoring shows that CREP sites have stable banks with adequate vegetation and no active erosion.
	Water Quality	Reduction of the rate of stream water heating to meet state ambient water quality standards by planting adequate vegetation on all riparian buffer lands.	• Annual monitoring shows that for small streams, CREP provides substantial canopy cover that should meet the goal of reduced stream water heating for salmonid health. • In several watersheds where CREP has been implemented along a large portion of the creek (Tucannon River in southeast Washington; Tenmile Creek in Whatcom County; Chimicum Creek in Jefferson County), stream temperatures have declined over time and salmon have returned in greater numbers than before CREP.
	Water Quality	Provision of a contributing mechanism for farmers and ranchers to meet the water quality requirements established under federal law and under Washington's water quality laws.	• Farmers and ranchers that participate in CREP immediately realize a physical separation of their activity from streams and salmon habitat. Provision of off-stream water when riparian

Agreement Name	Goal Type	Goal	Progress and Methods Used
			forest buffers are installed also enables ranchers to comply with water quality standards. • A study by the Washington State Department of Agriculture (Hancock, et.al., 2019) found that hedgerows, a practice allowed under Washington CREP on small streams, reduce pesticide deposition from aerial spray operations by 96 percent compared to fields without hedgerows.
	Habitat	Provision of adequate riparian buffers on 2,700 stream miles to permit natural restoration of stream hydraulic and geomorphic characteristics which meet habitat requirements of salmonids.	• Past and current CREP projects provide riparian forest buffers, hedgerows, filter strips, and wetland enhancements along over 900 miles of streamside on streams that support salmon. The annual increase during FY 2023 was 69.62 miles.
West Virginia	Program enrollment	Enroll 9,160 acres annually.	• West Virginia saw an increase of new and re- enrolled acres in FY 2023. There were 106.2 new acres establishing riparian buffers and protecting waterways. 452 acres were re-enrolled into CREP to continue maintaining an established buffer and continue protecting a waterway. • These new acres mean there are more sites throughout the State rehabilitating a waterway and establishing a stable buffer.
	Water Quality	Reduce runoff, sediment, and nutrients from agricultural enterprises.	• These goals are monitored through the State's WIP goals, EPA models, West Virginia Department of Environmental Protection (DEP) reviews, West Virginia Department of Agriculture (WVDA) monitoring, Nutrient Management Plans, EPA test

Agreement Name	Goal Type	Goal	Progress and Methods Used
			sites, local water districts, and Source Water Protection Plans (SWPP).
			• West Virginia saw an increase of new and re- enrolled acres for FY 2023. There were 106.2 new acres establishing a riparian buffer and protecting a waterway. 452 acres were re-enrolled into CREP to maintain established buffers and protect waterways.
	Water quality and wildlife habitat	Improvement of environmentally sensitive areas.	• Monitored through the State's WIP goals, EPA models, DEP reviews, WVDA monitoring, Nutrient • Management Plans, EPA test sites, local water districts, and Source Water Protection Plans (SWPP). • Water quality has been improved and wildlife habitat has been increased due to the implementation of CREP practices. Those practices are Establishment of Permanent Introduced Grasses and Legumes (CP1) 115.13 acres, Filter Strips (CP21) 267.87 acres, and Establishment of Riparian Buffers (CP22) on 4,194.32 acres.
Wisconsin	Water Quality	Riparian Project Area: The state seeks to improve the water quality of several water bodies that drain agricultural lands throughout the State of Wisconsin, through a reduction of sediment and the nutrient loading to these water bodies. Improving water quality through CREP will assist in improving water quality in the state's list of impaired waters (303d list). Wisconsin's CREP includes over 95 percent of the waters	• The state tracks enrollment, payments, landowner transfers, buyouts and environmental benefit progress information through a statewide CREP database maintained by the Wisconsin Department of Agriculture, Trade and Consumer Protection (DATCP). • Wisconsin had 3,542 active CREP contracts on 35,558.6 acres as of October 1, 2023. The state also continues to actively monitor over 5,000 acres of perpetual easements

Agreement Name	Goal Type	Goal	Progress and Methods Used
		included on the state's list of impaired waters that are impacted by agricultural activities. The specific goals for Wisconsin's CREP are to: - Reduce nutrient runoff of phosphorus (610,000 pounds) and nitrogen (305,000 pounds) by up to 10 percent and sediment (335,000 tons) in runoff by up to 15 percent from cropland and pastureland in the project areas. - Establish riparian buffers on up to 50 percent(about 3,700 miles) of the stream miles in the project area that drain farmland which currently is without adequate buffers.	that are no longer considered active by the FSA. During FY 2023, the state processed and paid incentives for 116 CREP contracts totaling 1,307 acres. New enrollments account for 44 of the contracts on 603 acres along with reenrollment of 72 existing contracts on 846 acres that expired in FY 2023. All the contracts paid in FY 2023 were 15-year agreements; however, one 4.22-acre easement in Dunn County was also established in FY 2023 but was paid by the State after September 30th. Enrollment at the end of FY 2023 saw 51 fewer 15-year agreements than at the end of the previous FY, resulting in a reduction of 189 acres of conservation practices on the ground in Wisconsin through CREP.
			• This is due to the rate of contracts expiring or terminating being greater than enrollment in FY 2023. • Wisconsin authorized $28 million in funding for CREP. In FY 2023, the State processed 310 CREP payments to landowners. As of October 1, 2023, a total of 13,171 incentive and cost share payments were paid to landowners for CREP totaling $21.9 million. In addition, counties reported they have spent a total of $4.5 million in non-federal staff and other costs to administer CREP. State and local costs for CREP total $26.3 million combined since the program began. • The majority (71 percent) of land enrolled in the Wisconsin

Agreement Name	Goal Type	Goal	Progress and Methods Used
			CREP is under one of the riparian practices. Filter strips (CP21) continue to be the majority (66 percent) of the riparian practices while riparian buffers (CP22) are the next most common (21 percent). Wetland restorations (CP23) under active CREP agreements or easements in the state comprise just over 10 percent of the acres. • Estimated annual reductions of agricultural runoff are calculated on a site-by-site basis by County Land Conservation Department staff using a DATCP environmental benefits report form. This form is submitted with each CREP application, and estimates are compiled by DATCP staff into the state CREP database. The 2023 estimated total annual reduction in phosphorus, nitrogen and sediment runoff, along with the length of stream or shoreline buffered, are:
			• Pounds of Phosphorus: 75,986 • Pounds of Nitrogen: 40,882 • Tons of Sediment: 37,706 • Miles of Stream or Shoreline Buffered: 654 miles State CREP enrollment totals are often slightly different than federal totals, as multiple CRP-1 contracts can be associated with a single state CREP contract and active CREP contracts for the state include both 15-year agreements and perpetual easements. Additionally, FSA records acres as soon as the contract is approved and shows as active in the federal system, while the state records

Agreement Name	Goal Type	Goal	Progress and Methods Used
			acres once the practice under which they are enrolled has been implemented. The state works annually with the FSA state office to cross reference state and federal CREP enrollment for each CRP sign-up to monitor and ensure all CRP-1s are accounted for that federal fiscal year.
	Habitat and Water Quality	Grassland Project Areas: The state seeks to enhance wildlife habitats for endangered grassland birds and other wildlife species in two grassland project areas in north central and south-central Wisconsin; the goal is to improve habitat for several grassland birds included under the Endangered Species Act. The specific goal of Wisconsin's CREP is to establish an additional 10 percent (15,000 acres) of grassland habitat acreage within the grassland project areas to increase populations of endangered or threatened grassland birds.	• Wisconsin CREP has two grassland project areas that can enroll up to 15,000 acres. While these regions are also eligible for the riparian CREP practices, there are several conservation practices unique to the grassland project areas. These practices include CP1 (permanent introduced grasses), CP2 (permanent native grasses), and CP25 (rare and declining habitat: prairie and oak savanna restoration). For Wisconsin CREP, these grassland practices must be placed adjacent to and within 1,000 feet of an eligible water body and be on steep slopes to address water quality issues as well as provide habitat for birds and pollinators. Grassland practices tend to be larger in acreage than the riparian practices; thus, while there are fewer contracts in a smaller eligible region, grassland practices make up 29 percent of the acres enrolled in CREP statewide. In total, 11,700 acres of grassland conservation practices are currently under an active CREP contract within the CREP grassland project areas.
			• The CREP southern grassland project area reached its

Agreement Name	Goal Type	Goal	Progress and Methods Used
			maximum allowable acres (10,000 acres) of grassland practice acres in 2016 and has remained at or near its cap each year since. The rare and declining habitat: prairie and oak savanna restoration practice (CP25) makes up the majority of grassland practices established in the southern grassland project area. The northern grassland project area has seen an increase in enrollment activity over the past year due to promotional efforts by the local county, FSA, and NRCS offices with an additional 122 acres of grassland practices enrolled. This area has room for enrollment with over half of the allowable acres still available. Permanent introduced grasses (CP1) is the primary practice established in the northern grassland region and most of the CREP contracts are signed up as 15-year agreements; however, there are 265 acres under CREP perpetual conservation easements.
	Water Quality	Reduce the peak runoff flow in critical areas of the Lake Superior Basin by 10 percent to 15 percent. The reduction in peak flow will be greater in the upper portions of watersheds with concentrations of agricultural lands. On individual participating farmland, reduce the overall hydrologic runoff curve numbers (RCN) from an average RCN of 89 to an RCN of 77 on the red clay soils on cropland and marginal pastureland. Based on models, this should	• The Lake Superior Project area provides the same conservation practice options for landowners to choose from as the riparian project area along with the addition of CP30 (Marginal Pastureland Wetland Buffer). Minimum and maximum widths for buffer practices in this region, however, are set higher to target reduction of peak flow rate as described in the project area goals. Incentive and annual rental payment incentives are also higher for practices in this project area

Agreement Name	Goal Type	Goal	Progress and Methods Used
			from both the state and federal government with state incentives double what they are in other project areas.
		reduce peak flow in streams, thereby reducing stream scouring (bank cutting and erosion) and sediment levels in the streams and to their Lake Superior outlets.	• The primary objective for CREP practices installed in this region is to reduce the peak channel flow in streams within critical areas of the Lake Superior Basin. Specifically on individual fields, CREP practices aim to reduce the hydrologic runoff curve number (RCN) on the red clay soils on cropland and marginal pastureland. According to the agreement application, CREP riparian buffers will greatly reduce the rate of runoff due to the "roughness" of the vegetation and the reduction in raindrop energy. Planting trees on slumping banks will not only stabilize the bank and reduce sediment loss to the stream, but the trees also will prevent small channels from becoming further incised. • Since the Lake Superior project area was authorized, two of the four counties in the area had landowners enroll in the CREP program. A total of five contracts covering 132.4 acres have been enrolled, with three perpetual easements on 43.1 acres. All the contracts were for riparian forest buffer (CP22) conservation practices. As of 9/30/2022, the two 15-year agreements covering 90.3 acres expired and were not reenrolled in FY 2023.

A successful hunt on CREP land in South Dakota.

Office of the Secretary Washington, DC 20250
May 23, 2024

The Honorable John Boozman
Ranking Member
Committee on Agriculture, Nutrition, and Forestry United States Senate
328A Russell Senate Office Building Washington, DC 20510

Dear Ranking Member Boozman:

This report to Congress is provided in accordance with Section 1231A of the Food Security Act of 1985, as amended, which requires the Secretary of Agriculture to submit an annual report to Congress that describes, with respect to each Conservation Reserve Enhancement Program (CREP) agreement, the: 1) status of the agreement; 2) purpose and objectives of the agreement; 3) Federal and partner commitments made under the agreement; and 4) progress made in fulfilling those commitments. This report contains fiscal year 2023 data and information.

If you have any questions, please have a member of your staff contact the Office of Congressional Relations at (202) 720-7095. A similar letter and the report are being sent to Chairwoman Debbie Stabenow.

Sincerely,
/s/ Thomas J. Vilsack THOMAS J. VILSACK
Secretary

Enclosure

Office of the Secretary Washington, DC 20250
May 23, 2024

The Honorable David Scott
Ranking Member Committee on Agriculture
U.S. House of Representatives
1010 Longworth House Office Building Washington, DC 20515

Dear Ranking Member Scott:

This report to Congress is provided in accordance with Section 1231A of the Food Security Act of 1985, as amended, which requires the Secretary of Agriculture to submit an annual report to Congress that describes, with respect to each Conservation Reserve Enhancement Program (CREP) agreement, the: 1) status of the agreement; 2) purpose and objectives of the agreement; 3) Federal and eligible partner commitments made under the agreement; and 4) progress made in fulfilling those commitments. This report contains fiscal year 2023 data and information.

If you have any questions, please have a member of your staff contact the Office of Congressional Relations at (202) 720-7095. A similar letter and the report are being sent to Ranking Member Glenn Thompson.

Sincerely,
/s/ Thomas J. Vilsack THOMAS J. VILSACK
Secretary

Enclosure

Office of the Secretary Washington, DC 20250
May 23, 2024

The Honorable Debbie Stabenow
Chairwoman
Committee on Agriculture, Nutrition, and Forestry United States Senate
328A Russell Senate Office Building Washington, DC 20510

Dear Chairwoman Stabenow:

This report to Congress is provided in accordance with Section 1231A of the Food Security Act of 1985, as amended, which requires the Secretary of Agriculture to submit an annual report to Congress that describes, with respect to each Conservation Reserve Enhancement Program (CREP) agreement, the: 1) status of the agreement; 2) purpose and objectives of the agreement; 3) Federal and partner commitments made under the agreement; and 4) progress made in fulfilling those commitments. This report contains fiscal year 2023 data and information.

If you have any questions, please have a member of your staff contact the Office of Congressional Relations at (202) 720-7095. A similar letter and the report are being sent to Ranking Member John Boozman.

Sincerely,
/s/ Thomas J. Vilsack THOMAS J. VILSACK
Secretary

Enclosure

Office of the Secretary Washington, DC 20250
May 23, 2024

The Honorable Glenn "GT" Thompson
Chairman

Committee on Agriculture
U.S. House of Representatives
1301 Longworth House Office Building Washington, DC 20515

Dear Chairman Thompson:

This report to Congress is provided in accordance with Section 1231A of the Food Security Act of 1985, as amended, which requires the Secretary of Agriculture to submit an annual report to Congress that describes, with respect to each Conservation Reserve Enhancement Program (CREP) agreement, the: 1) status of the agreement; 2) purpose and objectives of the agreement; 3) Federal and eligible partner commitments made under the agreement; and 4) progress made in fulfilling those commitments. This report contains fiscal year 2023 data and information.

If you have any questions, please have a member of your staff contact the Office of Congressional Relations at (202) 720-7095. A similar letter and the report are being sent to Ranking Member David Scott.

Sincerely,

/s/ Thomas J. Vilsack

Thomas J. Vilsack
Secretary

Enclosure

Chapter 4

Grassland Conservation Reserve Program (CRP): Signup 206 Ranking Factors*

U.S. Department of Agriculture

Overview

USDA's Farm Service Agency (FSA) Grassland Conservation Reserve Program (CRP) Ranking factors have been updated from previous signups. Grassland CRP is a working lands program which helps landowners and operators protect grassland, including rangeland, pastureland, and certain other lands, while maintaining the areas as grazing lands. The program emphasizes support for grazing operations, plant and animal biodiversity, and eligible land containing shrubs and forbs under the greatest threat of conversion. More information on grassland CRP signup is available in the FSA fact sheet, "Grassland Conservation Reserve Program – Fact Sheet."

Ranking Grassland

CRP Offers

FSA will rank offers for Grassland CRP using grassland ranking factors developed by the agency. FSA assigns each offer a point score based on the offer's relative ranking factors, with each offer competing against all other

---

* This is an edited, reformatted and augmented version of U.S. Department of Agriculture, Farm Service Publication, dated June 2024.

In: The Conservation Reserve Program
Editor: Jerry P. Norton
ISBN: 979-8-89530-774-8

offers. FSA determines offer acceptability based on the scoring results. For the current Grassland CRP batching FSA will use the following factors to determine acceptability of the land offered:

- Existence of expiring CRP;
- Existing Grassland;
- Existing Multi-species cover and predominance of native species;
- State Focus Area (land-based) determined in consultation with State Technical Committee;
- Applicant is an eligible Beginning, Veteran or Socially Disadvantaged farmer or rancher;
- Cost;
- Existing Small Livestock Operation;
- Other factors as determined by FSA

F-1. Current and Future Use: (0 - 30 Points)

Factor F1 encourages continued usage of the land and retains the benefits of expiring CRP or expiring GRP land using 2 sub factors. The formula for F1 = F1a + F1b.

- F1a – Expiring CRP or GRP: (0 to 20 points)

More Information

On grassland CRP signup is available in the FSA fact sheet, "Grassland Conservation Reserve Program – Fact Sheet."

For more information about FSA and its programs, visit fsa.usda. gov or your local FSA office.

To find your local FSA office, visit farmers.gov/ servicelocator.

- F1b – CRP Cover: (0 or 10 points)

This factor provides additional points depending on the current CRP cover. Assign 10 points if over 50 percent of the offered acreages is currently CP2, CP10 or CP25. Otherwise, assign 0 points.

F2 Beginning, Socially Disadvantaged, Limited Resource and Veteran Farmers and Ranchers: (0 to 20 Points)

For this factor assign 20 points if at least 50 percent of the CRP-1 shares are earned by a producer(s) meeting the requirements of the CCC-860.

Otherwise, assign 0 points.

Beginning farmers that served as a member of the armed forces can also check the "Veterans" block on the CRP-2G.

F3 Maximizing Grass- land Preservation: (0 to 35 Points)

Factor F3 targets land under the greatest threat of conversion, using 3 sub factors. The formula for F3 = F3a + F3b + F3c.

- F3a – Threat of Conversion: 0, 2.5, 5, 7.5 or 10 points

If over 50 percent of the offered acreage is in a county under threat of conversion as determined by FSA (0, 2.5, 5, 7.5 or 10 points).

- F3b – Threat of Conversion GIS: 0 or 20 points

If over 50 percent of the offered acreage is in the threat of conversion GIS Layer, assign 20 points (0 or 20 points).

- F3c – Longer Contract Length: 0 or 5 points

If the offer is for a 15-year contract, assign 5 points (0 or 5 points).

F4 Vegetative Cover: (0 to 30 Points)

This factor evaluates based on the covers on the acres offered. FSA or TSP shall only allow covers that are suitable for the site designated for the offer.

F4 values are determined by calculating the weighted average score for all practices in Table 1.

F5 Environmental Factors: (0 to 45 Points)

This factor is designed to help producers address acreages that are known to be in areas of designated candidate, threatened, or endangered species or critical habit or environmental sensitive areas under significant threat. The formula for F5 = F5a + F5b + F5c.

- F5a – Wildlife Priority Zones (0 or 15 points)

Grassland CRP will use the existing State wildlife priority zones identified for CRP general signup. For this factor assign 15 points if over 50 percent of the offered acreage is within an approved State wildlife priority zone. Otherwise, assign 0 points.

- F5b – State Grassland Priority Enrollment Criteria (0 or 15 points)

State Technical Committees submit geographic areas (whole counties) to address the specific conservation practices and the species of concern or specific habitat restoration, such as tall grass prairie, that would be addressed by Grassland CRP practice and/ or components to the FSA Conservation Division for approval. For this factor assign 15 points if over 50 percent of the offered acreage is in a State Grassland CRP zone. Otherwise, assign 0 points.

Table 1. Cover values

Practice CP88	F4 Points
Permanent Introduced Grasses and Legumes	
Monoculture stand.	5
Existing stand of 2 or more species of an introduced grass species.	10
Existing stand (minimum of 4 species) of at least 3 introduced grasses and at least 1 forb or legume of species best suited for wildlife in the area.	20
Permanent Native Grasses and Legumes	
Monoculture stand.	10
Existing stand (minimum of 3 species) of at least 2 native grass species.	20
Existing stand (minimum of 5 species) of at least 3 native grasses.	30

- F5c – National Priority Zone: (0 or 15 points)

This factor is designed to help producers address acreages that are environmental sensitive areas under significant threat. For this factor assign 15 points if over 50 percent of the offered acreage is in National Grassland CRP Priority Zone. Otherwise, assign 0 points.

F6 – Small Scale Livestock Operation: (0 to 10 Points)

This factor is designed to help producers' small-scale livestock operations. Assign 10 points if the producer certifies on the CRP-2G-1 that they meet eligibility as a Small- Scale Livestock operation. Otherwise, assign 0 points.

F7 – Cost: (0 to 25 Points)

Factor F7 is designed to optimize the environmental benefits per dollar for CRP rental payments. The maximum points for F7 is 25 points.

F7 – Offer Less Than Maximum Payment Rate (0 to 25 Points)

Offers submitted with a rental rate equal to the maximum payment rate will receive 0 points. Offers submitted below the maximum payment rate will receive points according to the corresponding value in Table 2.

Table 2. Offer less than the payment rate

Percent Below Maximum Payment Rate	F7 Points
1	2
2	4
3	6
4	8
5	10
6	12
7	14
8	16
9	18
10	20
11	21
12	22
13	23
14	24
>= 15	25

Ranking Factors Threshold for Acceptance

After the close of the CRP signup, FSA will rank all eligible offers based on their total point score for that offer.

The Secretary of Agriculture will then determine the ranking point threshold used to accept offers for the signup.

Making Grassland CRP Offers More Competitive

Offer competition is expected to be significant and only grassland CRP offers ranking high enough to be accepted will move forward toward enrollment. To make offers more competitive, consider all the ranking factors as well as the existing cover for the acreage offered. FSA also encourages producers to consult NRCS or authorized TSP regarding steps to take to maximize their offers ranking factor points to increase the likelihood that the offer will be accepted.

Chapter 5

Forest Management Incentive (FMI): Conservation Reserve Program*

U.S. Department of Agriculture

Overview

USDA Farm Service Agency's (FSA) Conservation Reserve Program (CRP) is a voluntary program that contracts with agricultural producers so that environmentally sensitive agricultural land is not farmed or ranched, but instead devoted to conservation benefits. CRP participants establish long-term, resource-conserving plant species, such as approved grasses or trees (known as "covers") to control soil erosion, improve water quality, and enhance wildlife habitat. In return, FSA provides participants with rental payments and cost- share assistance.

The Inflation Reduction Act of 2022 and Further Continuing Appropriations and Other Extensions Act, 2024 extended the authority and provided funding for of the Forestry Management Incentives Program (FMI) through September 30, 2024. On Nov. 16, 2023, President Biden signed into law, H.R. 6363, the Further Continuing Appropriations and Other Extensions Act, 2024, which extended the 2018 Farm Bill. The Inflation Reduction Act of 2022 (Pub. L. 117-169) and the Further Continuing Appropriations and Other Extensions Act, 2024 (Pub. L. 118-22), extended the authority and provided funding for the Forest Management Incentive (FMI).

---

* This is an edited, reformatted and augmented version of U.S. Dewpartment of Agriculture, Farm Service Agency Publication, dated January 2024.

In: The Conservation Reserve Program
Editor: Jerry P. Norton
ISBN: 979-8-89530-774-8

CRP FMI Signup

CRP FMI signup began on January 12, 2024, and is ongoing. Signup will continue for as long as CRP authority and CRP FMI funds are available.

Who is Eligible?

CRP FMI is limited to those farmers, ranchers, and forest landowners currently participating in CRP who have established conservation covers comprised of trees.

Eligible Land

All land currently enrolled in the following conservation practices is eligible for CRP FMI:

- CP3 – Tree Planting
- CP3A – Hardwood Tree Planting
- CP4D – Permanent Wildlife Habitat
- CP11 – Vegetative Cover Trees Already Established
- CP22 – Riparian Buffer
- CP22B – Denitrifying Bioreactor on Riparian Buffer
- CP22S – Saturated Riparian Buffer
- CP31 – Bottomland Timber Establishment on Wetlands
- CP36 – Longleaf Pine Establishment
- CP38A-22 – SAFE Riparian Buffer
- CP38C-3 – SAFE Tree Planting
- CP38C-3A – SAFE Hardwood Tree Planting
- CP38C-11 – SAFE Vegetative Cover Trees Already Established
- CP38C-31 – SAFE Bottomland Timber Establishment on Wetlands
- CP38D-36 – SAFE Longleaf Pine Establishment
- CP38E-4D – SAFE Permanent Wildlife Habitat

Eligible Technical Practices

The following technical practices are eligible for CRP FMI:

- 314 – Brush Management
- 315 – Herbaceous Weed
- Control
- 327 – Conservation Cover
- 338 – Prescribed Burning
- 394 – Firebreak
- 645 – Upland Wildlife
- Habitat
- 647 – Early Successional
- Habitat
- 660 – Tree/Shrub Pruning
- 666 – Forest Stand Improvement.

CRP FMI Payments

CRP FMI is a payment to eligible CRP participants who have properly completed authorized forest management practice activities, to improve condition of resources, promote forest management, or enhance wildlife habitat.

CRP FMI payments are equal to the lesser of the CRP participant's actual cost to implement the practice or 85% of the NRCS National Payment Scenario rate for the applicable NRCS technical practice.

A CRP participant may be eligible for more than one CRP FMI payment if multiple approved forest management activities are performed according to the conservation plan.

Approved forest management activities must be completed and reported by no later than the expiration date of the agreement, for the CRP participant to receive CRP FMI payment for such activity.

CRP FMI payment is conditional upon completion, and certification on the FSA-848B agreement, of approved forest management activities as provided in the conservation plan.

Total CRP FMI payment is limited to $200,000 per person or legal entity.

More Information

CRP is authorized by the Food Security Act of 1985 and was reauthorized through September 30, 2024, by Further Continuing Appropriations and Other Extensions Act, 2024 (Pub. L.118-22). The program is also governed by regulations published in 7 CFR, part 1410. The program is implemented by FSA on behalf of USDA's Commodity Credit Corporation.

This fact sheet is for informational purposes only; other restrictions may apply.

For more information, visit fsa.usda. gov or contact your local FSA office. To find your local FSA office, visit farmers.gov/service-locator.

Chapter 6

Farm Service Agency's: Conservation Reserve Program*

U.S. Department of Agriculture

Overview

USDA Farm Service Agency's (FSA) Conservation Reserve Program (CRP) is a voluntary program that contracts with agricultural producers so that environmentally sensitive agricultural land is not farmed or ranched, but instead devoted to conservation benefits. CRP participants establish long-term, resource- conserving plant species, such as approved grasses or trees (known as "covers") to control soil erosion, improve water quality and develop wildlife habitat. In return, FSA provides participants with rental payments and cost-share assistance. Contract duration is between 10 and 15 years.

CRP is authorized by the Food Security Act of 1985 and was reauthorized by the Agricultural Improvement Act of 2018 (the 2018 Farm Bill). The program is also governed by regulations published in 7 CFR, part 1410. The program is implemented by FSA on behalf of USDA's Commodity Credit Corporation.

* This is an edited, reformatted and augmented version of United States Department of Agriculture, Farm Service Agency Publication, dated February 2022.

In: The Conservation Reserve Program
Editor: Jerry P. Norton
ISBN: 979-8-89530-774-8

Program Administration

FSA administers CRP while technical support functions are provided by: USDA's Natural Resources Conservation Service (NRCS); State forestry agencies; Local soil and water conservation districts and; Non-federal technical assistance providers.

Benefits

CRP protects more than 20 millions of acres of American topsoil from erosion and is designed to safeguard the nation's natural resources. By reducing water runoff and sedimentation, CRP protects groundwater; helps improve the condition of lakes, rivers, ponds and streams; and is a major contributor to increased wildlife populations in many parts of the country.

USDA photo.

General Enrollment

Under general enrollment, producers have the opportunity to offer land for CRP general enrollment annually during announced enrollment periods.

Offers for CRP contracts are ranked according to the Environmental Benefits Index (EBI). FSA uses the following EBI factors to assess the environmental benefits for the land offered:

- Wildlife habitat benefits resulting from covers on
- contract acres;
- Water quality benefits from reduced erosion, runoff and
- leaching;
- On-farm benefits from reduced erosion;
- Benefits that will likely endure beyond the contract period;
- Air quality benefits from reduced wind erosion and;
- Cost.

Further information on CRP general signup is available in the FSA fact sheet "Conservation Reserve Program General Signup."

Continuous Enrollment

Under CRP continuous enrollment, environmentally sensitive land devoted to certain conservation practices may be enrolled in CRP at any time. Certain eligibility requirements still apply, but offers are not subject to competitive bidding. Further information on CRP continuous enrollments is available in the FSA fact sheet *"Conservation Reserve Program Continuous Enrollment."*

Grasslands Enrollment

CRP Grasslands helps landowners and operators protect grassland, including rangeland, and pastureland, and certain other lands, while maintaining the areas as grazing lands. The program emphasizes support for grazing operations, plant and animal biodiversity, and grassland and land containing shrubs and forbs under the greatest threat of conversion. Further information on CRP Grassland enrollments is available in the FSA fact sheet *"Conservation Reserve Program Grasslands."*

Who is Eligible?

A producer must have owned or operated the land for at least 12 months prior to submitting the offer for continuous or 12 months before the close of general or grasslands signup, unless:

- The new owner acquired the land due to the previous owner's death;
- The ownership change occurred due to foreclosure where the owner exercised a timely right of redemption in accordance with state law or;
- The circumstances of the acquisition present adequate assurance to FSA that the new owner did not acquire the land for the purpose of placing it in CRP.

Eligible Land

For cropland, land must be planted or considered planted to an agricultural commodity for four of six crop years from 2012 to 2017, and that is physically and legally capable of being planted (no planting restrictions due to an easement or other legally binding instrument) in a normal manner to an agricultural commodity.

For general signup, land also must meet one of the following criteria:

- Have a weighted average erosion index of eight or
- higher;
- Be enrolled in a CRP contract that expires Sept. 30 or;
- Be located in a national or state CRP conservation priority area.

Marginal pasture land may also be eligible for continuous signup.
CRP Grasslands has alternative requirements for eligible land.

Payments

FSA provides CRP participants with annual rental payments, as well as certain incentive payments and cost- share assistance.

In return for establishing long-term, resource- conserving covers, FSA provides annual rental payments to participants. FSA bases rental rates on the relative productivity of the soils within each county and the average cash rent using data provided by the National Agricultural Statistics Service (NASS). The soil rental rates are subject to an 85 percent proration for general signup and a 90 percent proration for continuous signup. The maximum CRP rental rate for each offer is calculated in advance of enrollment. Producers may offer land at that rate or offer a lower rental rate.

CRP grasslands rates are based on 75 percent of the NASS pasture rate.

More Information

This fact sheet is provided for informational purposes only; other restrictions may apply. For details, contact your local FSA office. To find your local FSA office, contact your local service center and USDA Farm Service Agency office: *farmers.gov/service-locator*.

Chapter 7

USDA Accepts More Than 2.2 million Acres Through 2024 Grassland, General and Continuous Conservation Reserve Program Offers*

U.S. Department of Agriculture

1400 Independence Ave. SW
Washington, DC 20250
FPAC.BC.Press@usda.gov

Enrollment Nears the Conservation Reserve Program 27 Million Acre Cap

WASHINGTON, Aug. 19, 2024 – The U.S. Department of Agriculture (USDA) is accepting offers for more than 2.2 million acres from agricultural producers and private landowners through this year's Grassland, General, and Continuous Conservation Reserve Program (CRP) signups administered by USDA's Farm Service Agency (FSA). With these accepted acres, enrollment is very near the 27 million CRP acreage cap.

"The popularity of our suite of Conservation Reserve Programs, including the significant success of Grassland Conservation Reserve Program and the fact that we are nearing the Conservation Reserve Program acreage cap, is a testament to the commitments made by the farmers, ranchers and landowners we serve to mitigate the impacts of climate change by conserving, protecting and enhancing our natural resources," said Zach Ducheneaux, Administrator

* This is an edited, reformatted and augmented version of U.S. Department of Agriculture Publication.

In: The Conservation Reserve Program
Editor: Jerry P. Norton
ISBN: 979-8-89530-774-8

of USDA's Farm Service Agency (FSA). "It's through all FSA's conservation programs that we are able to work directly with agricultural producers to implement climate smart conservation practices that ultimately benefit everyone."

Grassland CRP

Including the nearly 1.44 million acres recently accepted in Grassland CRP for offers received in 2024, producers will have enrolled more than 10 million acres in this popular working-lands program.

Grassland CRP allows producers and landowners to continue grazing and haying practices while protecting grasslands and further CRP's impacts. Grassland CRP leverages working lands practices to improve biodiversity and conserve environmentally sensitive land.

Enrolled acres help sequester carbon in vegetation and soil, while enhancing resilience to drought and wildfire. Meanwhile, producers can still conduct common grazing practices, such as haying, mowing or harvesting seed from the enrolled land, which supports agricultural production.

Top states for this year's Grassland CRP signup include:

- Nebraska with 237,853 acres accepted
- Colorado with 218,145 acres accepted
- New Mexico with 185,619 acres accepted

Additionally, to target conservation in key geographies, USDA prioritizes land within two National Priority Zones: The Greater Yellowstone Ecosystem and the Dust Bowl area. The 2024 Grassland CRP enrollment has added more than 560,000 acres in the Priority Zones, bringing the total acres to 3.8 million. By conserving working grasslands and other lands that underpin iconic big game migrations, land enrolled in these zones contributes to broader USDA conservation efforts through Working Lands for Wildlife.

General and Continuous CRP

FSA also has accepted nearly 200,000 acres through the General signup, bringing total acres enrolled in General CRP to 7.9 million acres. As one of the largest private lands conservation programs in the United States, CRP offers a range of conservation options to farmers, ranchers and landowners. It has been an especially strong opportunity for farmers with less productive or marginal cropland, helping them re-establish valuable land cover to help improve water quality, prevent soil erosion and support wildlife habitat.

In January 2024, FSA opened enrollment for Continuous CRP. Under this enrollment, producers and landowners can enroll in CRP throughout the year. Offers are automatically accepted provided the producer and land meet the eligibility requirements and the enrollment levels do not exceed the statutory cap. To date, in 2024, more than 565,000 acres have been offered through Continuous CRP bringing the total acres enrolled to 8.5 million.

Producers can still make an offer to participate in CRP through the Continuous CRP signup, which is ongoing, by contacting FSA at their local USDA Service Center.

More Information

FSA helps America's farmers, ranchers and forest landowners invest in, improve, protect and expand their agricultural operations through the delivery of agricultural programs for all Americans. FSA implements agricultural policy, administers credit and loan programs, and manages conservation, commodity, disaster recovery and marketing programs through a national network of state and county offices and locally elected county committees. For more information, visit, fsa.usda.gov.

USDA touches the lives of all Americans each day in so many positive ways. Under the Biden-Harris administration, USDA is transforming America's food system with a greater focus on more resilient local and regional food production, fairer markets for all producers, ensuring access to safe, healthy and nutritious food in all communities, building new markets and streams of income for farmers and producers using climate smart food and forestry practices, making historic investments in infrastructure and clean energy capabilities in rural America, and committing to equity across the Department by removing systemic barriers and building a workforce more representative of America. To learn more, visit usda.gov.

Chapter 8

Conservation Reserve Program: Improving How USDA Selects Land Could Increase Environmental Benefits*

United States Government Accountability Office

Abbreviations

CREP	Conservation Reserve Enhancement Program
CRP	Conservation Reserve Program
EBI	Environmental Benefits Index
ERS	Economic Research Service
FSA	Farm Service Agency
FWS	U.S. Fish and Wildlife Service
FY	fiscal year
NASS	National Agricultural Statistics Service
NRCS	Natural Resources Conservation Service
SAFE	State Acres for Wildlife Enhancement
USDA	U.S. Department of Agriculture

* This is an edited, reformatted and augmented version of the United States Government Accountability Office Report to Congressional Requesters, Publication No. GAO-24-106311, dated September 2024.

In: The Conservation Reserve Program
Editor: Jerry P. Norton
ISBN: 979-8-89530-774-8

Why GAO Did This Study

CRP is one of the largest USDA conservation programs for agricultural land in the U.S. In fiscal year 2023, USDA provided $1.8 billion in rental payments for 22.9 million acres enrolled in CRP. CRP, which FSA administers, was established by the Food Security Act of 1985 and most recently reauthorized by the Agriculture Improvement Act of 2018. It has three types of enrollment—general, which includes land that may have at least a certain level of soil erosion; continuous, which focuses on small, targeted portions of land; and grassland, which is typically used for livestock grazing. CRP's implementing regulation cites cost-effectiveness as one objective.

GAO was asked to review FSA's oversight of CRP. This report, among other things, examines the methods FSA uses to select CRP offers and the extent to which FSA reviews these methods and updates them as needed.

To conduct this review, GAO analyzed USDA documents and interviewed eight experts, including current and former USDA officials and academics who authored studies on CRP. GAO also analyzed USDA data on CRP for fiscal years 1986 through 2023.

What GAO Recommends

GAO is making four recommendations, including that FSA develop a process to periodically review its methods for selecting CRP offers, and then address any resulting findings, as appropriate. USDA generally agreed with all four recommendations.

What GAO Found

To participate in the U.S. Department of Agriculture's (USDA) Conservation Reserve Program (CRP), landowners and operators take agricultural land out of use and implement conservation practices, or agree to protect grassland, in exchange for annual rental payments. CRP focuses on environmentally sensitive land, such as those on which crop production can result in environmental damage. The program has three types of enrollments—general, continuous, and grassland—that all share some grass-related conservation practices. Continuous enrollment can also include water quality practices.

To participate in general and grassland enrollments, landowners and operators make competitive offers, which USDA's Farm Service Agency (FSA) scores and selects for acceptance. To score these offers, FSA considers their potential benefits and costs. For example, for general enrollment, FSA calculates an offer's total score by combining points for environmental benefits related to soil erosion, wildlife habitat, and water quality, among other things, with points related to the relative cost. FSA then ranks and selects offers based on their scores. For continuous enrollment, FSA typically automatically accepts offers if they meet certain criteria, except for one project on migratory birds, butterflies, and pollinators, in which offers are competitive.

Examples of Land in USDA's Conservation Reserve Program

Prairie with wildlife Filter strip on farmland Cattle on grassland

Source: U.S. Department of Agriculture (USDA). | GAO-24-106311.

According to some studies GAO reviewed and experts GAO interviewed, FSA has opportunities to further improve its selection methods to increase the program's environmental benefits and cost-effectiveness. For example, FSA could increase points for establishing native wildflowers that benefit pollinators and expand the use of competition for continuous enrollment. FSA officials agreed that such opportunities could exist.

While FSA has generally made some small changes over the years to its selection methods, it does not have a systematic process to periodically review its methods. In 2023, FSA made a larger change by updating how it estimates soil erosion benefits to increase accuracy. FSA has held some meetings internally, such as after general enrollment periods, and with other federal agencies (most recently in 2019) to discuss its methods for selecting offers.

However, developing a process to regularly review its methods would better position FSA to identify potential improvements. Addressing any findings resulting from these periodic reviews could help increase CRP's cost-effectiveness and environmental benefits.

September 25, 2024

The Honorable Kirsten Gillibrand
Chair
Subcommittee on Livestock, Dairy, Poultry,
Local Food Systems, and Food Safety and Security Committee
on Agriculture, Nutrition, and Forestry
United States Senate

The Honorable Cory Booker
United States Senate

The Conservation Reserve Program (CRP) is a federal program that seeks to offset the detrimental effects that agricultural practices can have on the environment, such as soil erosion and runoff from fertilizers or pesticides entering waterways. The program also seeks to conserve and improve grassland. CRP pays landowners and operators to take environmentally sensitive agricultural land out of production and devote them to conservation practices.[1] These practices, which include establishing grasses or trees as land cover, can achieve long-term program goals such as preventing soil erosion, improving water quality, and reducing loss of wildlife habitat. For example, one study estimated that CRP increased the number of young waterfowl by 1.5 million annually in one prairie region.[2] For grassland, CRP pays landowners and operators to protect grassland, rather than taking the land out of use.

The Farm Service Agency (FSA), within the U.S. Department of Agriculture (USDA), administers CRP. To participate in CRP, landowners and operators offer land to FSA for conservation in exchange for rental payments.[3] FSA uses various methods to select from these offers, such as a competitive offer process that considers environmental benefits andcost. In fiscal year

---

[1] Operators include those who lease the land or otherwise have the landowner's approval to enroll in the program. Land on which crop production can result in environmental damage is considered environmentally sensitive.

[2] R. G. Drum et al., *Assessing the Biological Benefits of the USDA-Conservation Reserve Program (CRP) for Waterfowl and Grassland Passerines in the Prairie Pothole Region of the United States: Spatial analyses for targeting CRP to maximize benefits for migratory birds,* Final Report for USDA-FSA Agreement: 12-IA-MRE-CRP-TA (April 2015).

[3] To participate in the program, landowners and operators agree to devote whole or partial fields to conservation practices in exchange for an annual rental payment, incentive payments, and cost-share assistance from USDA.

(FY) 2023, USDA provided $1.8 billion in rental payments to landowners and operators for 22.9 million acres enrolled in the program. CRP is one of USDA's largest conservation programs, according to USDA's FY 2025 budget document.[4]

CRP was established by the Food Security Act of 1985 (1985 Farm Bill) and has been reauthorized multiple times, most recently in 2018.[5] The 1985 Farm Bill and the reauthorizing legislation set certain minimum or maximum numbers of acres that USDA can enroll in CRP.[6] Other factors, such as agricultural commodity prices and rental rates paid to landowners and operators, influence the amount of acreage enrolled in CRP, according to USDA and other studies.[7] The program's implementing regulation cites cost-effectiveness as part of the program's objectives.[8] Over the past 20 years, peer-reviewed studies by academia and USDA have identified opportunities for the program to more cost-effectively achieve environmental benefits, such as by adjusting its methods for selecting offers.

We last reported on CRP in 1995.[9] In that report, we found that CRP could provide longer-term environmental benefits at less cost with certain modifications. We recommended, for example, that Congress consider modifying CRP to focus more on creating buffer zones—strips of trees or other plants bordering waterbodies to improve water quality and wildlife

---

[4] Other large USDA programs, administered by USDA's Natural Resources Conservation Service, include the Environmental Quality Incentives Program and the Conservation Stewardship Program. Both of these programs work with producers to develop a conservation plan outlining conservation practices and activities based on management objectives for the farm or ranch.

[5] Food Security Act of 1985, Pub. L. No. 99-198, § 1231, 99 Stat. 1354, 1509; Agriculture Improvement Act of 2018, Pub. L. No. 115-334, § 2201(a), 132 Stat. 4490, 4530.

[6] The 1985 Farm Bill set the highest acreage limit at 40 million to 45 million acres by 1990. The 2014 Farm Bill set the lowest acreage limit at 24 million acres. The 2018 Farm Bill set the acreage limits so that it gradually increased from 24 million in FY 2019 to 27 million acres in FY 2023.

[7] Brian Cornish et al., "Impact of changes in Title II of the 2018 Farm Bill on the acreage and environmental benefits of Conservation Reserve Program," Applied Economic Perspectives and Policy, vol. 44, no. 2 (2022): 1100–1122. Heesun Jang et al., "An empirical structural model of productivity and Conservation Reserve Program participation," Land Economics, vol. 94, no. 1 (2018): 1–18. U.S. Department of Agriculture, Economic Research Service, The Influence of Rising Commodity Prices on the Conservation Reserve Program, Economic Research Report No. 110 (Washington, D.C.: February 2011).

[8] 7 C.F.R. § 1410.3(c).

[9] GAO, Conservation Reserve Program: Alternatives Are Available for Managing Environmentally Sensitive Cropland, RCED-95-42 (Washington, D.C.: Feb. 21, 1995).

habitat—rather than on removing whole fields from production. Congress adopted this recommendation in the 1996 Farm Bill.[10]

You asked us to review FSA's oversight of CRP. This report (1) describes the methods FSA uses to select CRP offers and the composition of acres and rental payments in the program, (2) examines the extent to which FSA reviews its selection methods and updates them as needed, and (3) examines the extent to which FSA analyzes and reports on CRP's environmental benefits. This report also provides data on various aspects of the program, such as average USDA rental payments per acre, the length of time that acres are enrolled in the program, the distribution of acres and rental payments by state and county, and the use of various conservation practices (see appendixes I through VI).

To provide information on the methods FSA uses to select offers, we focused on factors that FSA uses to select offers for three types of CRP enrollment, and the composition of acres and rental payments in CRP. We reviewed FSA documents such as FSA's handbook for the program.[11] We also reviewed FSA documents on the methods FSA uses to score, rank, and select offers (referred to as indexes) and determine the bid cap—the maximum rental rate that landowners and operators can receive for an offer. We also interviewed officials from FSA about these selection methods.

To examine the extent to which FSA reviews its methods to select offers, we focused on the benefits of FSA's methods for selecting land offered for CRP, according to studies and experts; the extent to which FSA uses science-based information to select CRP offers;[12] opportunities FSA has to improve its methods, according to studies and experts; and the extent to which FSA reviews its selection methods to determine whether they need updating. We reviewed FSA documents and interviewed officials.

We also compared FSA's actions against criteria, such as USDA's Fiscal Years 2022–2026 Strategic Plan and our key practices for evidence- based policymaking.[13]

---

[10] Federal Agriculture Improvement and Reform Act of 1996, Pub. L. No. 104-127, § 332, 110 Stat. 888, 994.

[11] U.S. Department of Agriculture, *Agricultural Resource Conservation Program,* 2-CRP (Washington, D.C.: Feb. 10, 2023).

[12] For the purpose of this report, we define science-based information as information based on scientific research and data.

[13] U.S. Department of Agriculture, Strategic Plan: Fiscal Years 2022–2026 (Washington, D.C.: March 2022) and GAO, Evidence-Based Policymaking: Practices to Help Manage and Assess the Results of Federal Efforts, GAO-23-105460 (Washington, D.C.: July 12, 2023). The key practices we used from this report are assessing the evidence, using evidence to learn, applying learning to decision-making, communicating learning and results, and involving stakeholders.

In addition, we reviewed agency and academic studies—including peer-reviewed academic journal articles and USDA reports—on CRP and FSA's methods for selecting CRP offers. We also interviewed eight experts identified through literature searches who primarily authored these reports and articles. See appendix VII for more information on how we selected these studies and experts.

To examine the extent to which FSA analyzes and reports on CRP's environmental benefits, we focused on the extent to which FSA has used science-based information to estimate key environmental benefits, and the extent to which FSA has analyzed and reported on environmental benefits by type of enrollment. We reviewed research that USDA conducted or funded related to CRP benefits.

To provide data on various aspects of CRP, we analyzed USDA data from FY 1986 through FY 2023, including on number of acres enrolled, USDA rental payments, and how those acres and rental payments are distributed by state and county. We reviewed the reliability of these data, including identifying any omissions and anomalies, and determined that they were sufficiently reliable for the purposes of reporting information on CRP acreage and payments. For more information on the scope and methodology of our data analysis, see appendix VIII.

We conducted this performance audit from October 2022 to September 2024 in accordance with generally accepted government auditing standards. Those standards require that we plan and perform the audit to obtain sufficient, appropriate evidence to provide a reasonable basis for our findings and conclusions based on our audit objectives. We believe that the evidence obtained provides a reasonable basis for our findings and conclusions based on our audit objectives.

Background

According to USDA's handbook for the program, CRP is intended to cost-effectively assist landowners and operators in conserving and improving the nation's natural resources, including by focusing on:

- protecting the nation's soil, water, and wildlife resources;
- improving and preserving water quality; and
- enhancing fish and wildlife habitat.

Landowners and operators participating in CRP use conservation practices such as establishing or maintaining native and introduced grasses and legumes, permanent wildlife habitats, and filter strips and restoring wetland.[14] Figure 1 shows examples of different types of land in CRP.

Source: U.S. Department of Agriculture (USDA). | GAO-24-106311.

Figure 1. Examples of Land in USDA's Conservation Reserve Program.

CRP was implemented in 1986 with one type of enrollment—general enrollment—but has evolved to encompass three types of enrollment:

- General enrollment. This type of enrollment is open to land that meet certain criteria, such as having at least a certain level of soil erosion.[15] Landowners and operators compete nationally for acceptance into CRP based, in part, on anticipated environmental benefits during specified enrollment periods. Land accepted for general enrollment is removed from agricultural production, and landowners and operators implement common conservation practices that can include establishing or maintaining vegetative cover of introduced grasses and legumes.
- Continuous enrollment. This type of enrollment focuses on small, targeted portions of land. Landowners and operators typically may enroll their land at any time (i.e., continuously) during the year

[14] Introduced species are those that evolved elsewhere and have been transported and purposely or accidentally disseminated by humans. A filter strip is a narrow band of grasses, legumes, and forbs—plants that are not woody and have broad leaves. Filter strips improve water quality by intercepting sediment and nutrients, mitigate erosion by reducing the negative impacts of wind and water, and provide habitat and corridors for wildlife. Wetlands are low-lying ecosystems where the water table is always at or near the surface. They perform vital ecological functions, including providing critical habitat for wildlife and waterfowl, mitigating floods by slowing down and absorbing excess water during storms, purifying water by filtering out pollutants before they enter streams and lakes, among other things.

[15] USDA uses an erodibility index, which includes the potential for water or wind erosion to adversely affect the long-term production of the offered land.

without competition if the land and conservation practices meet certain criteria. Land in continuous enrollment is removed from agricultural production, and the conservation practices implemented include those related to water quality, such as establishing filter strips and riparian buffers, and restoration of wetlands.[16] Continuous enrollment also shares some grass- related conservation practices with general enrollment.

- Grassland enrollment. This type of enrollment focuses on "working lands," or land that often continues to be used for livestock grazing while in CRP. The enrollment process for grassland is similar to general enrollment. In contrast to the other two types of enrollment, landowners and operators typically continue grazing operations while conserving and improving grassland, rather than taking the land out of use. Grassland enrollment tends to have the largest average number of acres per contract, according to USDA documents.

Table 1 provides information on the characteristics of each type of enrollment.

Table 1. Characteristics of enrollment types in USDA's Conservation Reserve Program (CRP)

	General	Continuous	Grassland
First year of enrollments	1986	1997	2016
Contract length (years)	10–15[a]	10–15[a,b]	10–15[a]
Crop history requirement	Yes[c]	Yes[c]	No
Land use while in program	Temporarily removed from production	Temporarily removed from production	Typically working land (haying/grazing)
Minimum acreage set by 2018 Farm Bill	None	8.0–8.6 million acres[d]	1.0–2.0 million acres[e]
Maximum acreage set by 2018 Farm Bill		24.0–27.0 million acres for the entire CRP[f]	

Source: GAO analysis of U.S. Department of Agriculture (USDA) information. | GAO-24-106311.

[a] The contract length for each type of enrollment is generally 10 to 15 years, but contract lengths for a certain pilot—the Soil Health and Income Protection Program—are 3 to 5 years.

[16] Riparian buffers are strips of trees bordering streams, wetland areas, and other waterbodies to improve water quality and wildlife habitat.

[b] A pilot within continuous enrollment—Clean Lakes, Estuaries, and Rivers Initiative—offers 30-year contracts.

[c] Federal law requires that, to be eligible for general or continuous enrollment, cropland must, in general, have been annually planted or considered planted to an agricultural commodity in 4 of the 6 crop years from 2012 through 2017.

[d] The 2018 Farm Bill sets a minimum continuous enrollment at 8.0 million acres for fiscal year (FY) 2019, 8.25 million acres for FY 2020, 8.5 million acres for FY 2021, and 8.6 million acres for FY 2022 and FY 2023. Agriculture Improvement Act of 2018, Pub. L. No. 115-334, § 2201(c)(3), 132 Stat. 4490, 4534 (codified at 7 U.S.C. § 3831(d)(6)(B)).

[e] The 2018 Farm Bill sets a minimum grassland enrollment at 1.0 million acres for FY 2019, 1.5 million acres for FY 2020, 2.0 million acres for FY 2021 through FY 2023. Pub. L. No. 115-334, § 2201(c)(2), 132 Stat. 4490, 4532 (codified at 7 U.S.C. § 3831(d)(2)).

[f] The 2018 Farm Bill set the maximum acreage enrollment for the entire CRP at 24.0 million acres in FY 2019, 24.5 million acres in FY 2020, 25.0 million acres in FY 2021, 25.5 million acres in FY 2022, and 27.0 million acres in FY 2023. Pub. L. No. 115-334, § 2201(c)(1), 132 Stat. 4490, 4531 (codified at 7 U.S.C. § 3831(d)(1)).

According to FSA's handbook for CRP, the process for general and grassland enrollments starts with the announcement of a sign-up period (see fig. 2).

- *USDA announces sign-up period.* USDA opens a sign-up period for general and grassland enrollments and invites landowners and operators to submit offers.
- *Landowners and operators notify FSA of their interest in enrolling in CRP. FSA provides details on how offers will be scored.* Landowners and operators inform FSA about the specific land they want to offer for CRP. FSA then provides landowners and operators with details on factors FSA will use to score and rank offers. FSA also provides landowners and operators with a maximum allowable rental rate that is specific to the offered land, as described in more detail below.
- *Landowners and operators submit offers for enrollment to FSA.* During the sign-up period, landowners and operators submit offers that include the conservation practices they propose to establish on their land.

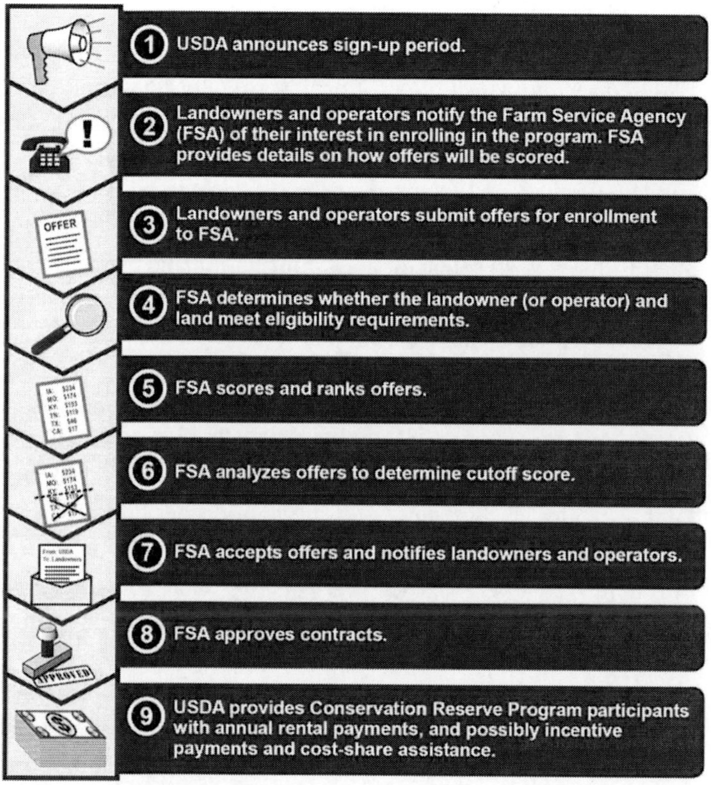

Sorce: GAO analysis of U.S. Department of Agriculture (USDA) information, GAO (icons). | GAO- GAO-24-106311.

Note: Operators include those who lease the land or otherwise have the landowner's approval to enroll in the program.

Figure 2. Process for General and Grassland Enrollments in USDA's Conservation Reserve Program.

- *FSA determines whether the landowner (or operator) and land meet eligibility requirements.* General and grassland enrollments share some landowner eligibility requirements. For example, a landowner or operator must have owned or operated the land for at least 12 months prior to the close of the sign-up period, or 12 months prior to submitting an offer under continuous sign-up periods. Each enrollment type also has additional land eligibility requirements specific to that type. For example, under general enrollment, the cropland must, in general, have been annually planted or considered planted to an agricultural commodity in 4 of the 6 crop years from

2012 through 2017.[17] Also, for grassland enrollment, the land must contain forbs or shrubland for which grazing is the predominant use.[18]

- *FSA scores and ranks offers.* For general and grassland enrollment, FSA uses an index that assigns points for various environmental benefits, as well as for a cost factor that gives fewer points for offers with higher bids (i.e., rental rates). Points for some environmental benefits in the general enrollment index are based on computer models that rely on science-based information and produce measures. For example, one model projects the amount of soil erosion that may be prevented for the land associated with a specific offer.[19] FSA staff determine points for other environmental benefits according to criteria in FSA's CRP handbook, which was developed using FSA and Economic Research Service (ERS) officials' professional judgment. USDA's information systems use the indexes to produce a total score for each offer. After the sign-up period ends, FSA uses these scores to rank all offers for general or grassland enrollment.
- *FSA analyzes offers to determine cutoff score.* Officials select a minimum score for each enrollment type that is based on the offers FSA received for that type.
- *FSA accepts offers and notifies landowners and operators.* FSA accepts all offers with a score above the cutoff score and notifies landowners and operators of acceptance. Landowners and operators with acceptable offers are able to withdraw their offers within a certain period of time.
- *FSA approves contracts.* FSA approves contracts for landowners and operators whose offers were accepted and who did not withdraw their offer to enroll in the program.

The process for landowners and operators to enroll in the program begins early in the year with USDA's announcement for landowners and operators to submit offers and ends no later than September 30 with approval of contracts. These contracts take effect and USDA begins making rental payments on October 1.

---

[17] 16 U.S.C. § 3831(b)(1)(B). Historically, this time period for land to have been annually planted or considered planted to an agricultural commodity is updated with each new farm bill.

[18] 16 U.S.C. § 3831(b)(3)(C). There are additional land eligibility requirements specific to general and grassland enrollment.

[19] Models are representations of real-world systems.

For all three types of enrollment, FSA determines a maximum rental rate that landowners and operators can receive, known as the bid cap. The bid cap is intended to avoid excessive payments. The cap is specific to the offered land and the type of enrollment and may include some adjustment factors.

- General and continuous enrollment. FSA determines bid caps for both types of enrollment by using data from the National Agricultural Statistics Service (NASS) on county average rent and the productivity of soils.[20] USDA adjusts the respective county's average rental rate according to factors such as the land's soil productivity.[21] For example, for land offered in a county with an average annual rental rate per acre of $100 and that has a productivity index of 1.5, with no additional adjustments, the bid cap would be $150. FSA then includes other factors in these rates, such as a 3, 5, or 10 percent incentive for CRP practices that will increase carbon sequestration or reduce greenhouse gas emissions, and a one-time 10 percent inflationary adjustment that is applied during the first year of enrollment.[22] After considering the various adjustments, the rental rate cannot exceed $240 per acre for general enrollment or $300 per acre for continuous enrollment, as of FY 2023.[23]
- Grassland enrollment. FSA sets the bid cap at 75 percent of the land's grazing value, which is determined by the NASS pasture rate, and does not provide any adjustments.

Besides the rental payments, which represent about 95 percent of total CRP payments, according to FSA officials, FSA provides other types of

---

[20] USDA's Cash Rents Survey provides the basis for county estimates of the cash rent paid for irrigated cropland, non-irrigated cropland, and pasture. The 2008 Farm Bill mandated that NASS provide mean rental rates for all counties with 20,000 or more acres of cropland plus pasture. NASS conducts the county-level Cash Rents Survey every year in all states except Alaska and releases national and state estimates each August, according to NASS's website. For more information, see https://www.nass.usda.gov/Surveys/.

[21] The soil productivity factor is measured by the soil productivity index, which is based on the publicly available National Commodity Crop Productivity Index developed by NRCS. CRP enrollment has generally been capped at 25 percent of the cropland in each county.

[22] This includes a 20 percent rental rate incentive for CRP water quality practices that will improve water quality by reducing sediment loading, nutrient loading, and harmful algal blooms, according to an FSA document.

[23] An exception to this limit within CRP is the Conservation Reserve Enhancement Program (CREP). FSA's CREP is a public-private partnership program that allows states, tribal governments, non-profits, and private entities to partner with FSA to implement CRP practices that address certain conservation and environmental objectives.

financial incentives for landowners and operators to offer land for enrollment in CRP. These include cost-share assistance and incentive payments. For example, FSA provides a cost-share of up to 50 percent of the reimbursable cost of installing a conservation practice. For continuous enrollment only, FSA also provides incentives and adjustments such as a sign-up incentive payment of 32.5 percent of the first full year's annual rental payment at contract approval.

FSA administers CRP with support from other USDA agencies. Specifically, USDA's Natural Resources Conservation Service (NRCS), which administers other USDA conservation programs, provides technical support functions.[24] Other USDA agencies such as the Agricultural Research Service, ERS, the National Institute of Food and Agriculture, and the U.S. Forest Service provide research and educational and technical assistance.

USDA Uses Various Methods to Select CRP Offers for Each Type of Enrollment

USDA's FSA uses various methods to select CRP offers for each type of enrollment. It uses similar competitive methods to select offers for general and grassland enrollments. Specifically, for general enrollment, FSA uses an Environmental Benefits Index (EBI) that is composed of factors representing environmental benefits and a factor related to cost. For grassland enrollment, FSA uses an index composed of benefits factors and a factor related to cost. For continuous enrollment, FSA typically automatically accepts offers that meet certain criteria.

What Factors Does FSA Use to Select Offers for General Enrollment?

To select offers for general enrollment, FSA uses the EBI, a competitive method to score offers. The EBI has six ranking factors that total to determine the EBI score. Five factors relate to environmental benefits— soil erosion, water quality, air quality, wildlife, and enduring benefits— while the sixth

---

[24] FSA and NRCS currently administer over 20 programs and subprograms that are directly or indirectly available to assist producers and landowners who wish to practice conservation on agricultural land.

relates to the cost per acre of the offer (see fig. 3).[25] Cost factor points are calculated in part based on the extent to which proposed rental rates are below the maximum allowed rental rates (bid caps). Lower rental rates receive more points than higher rental rates.

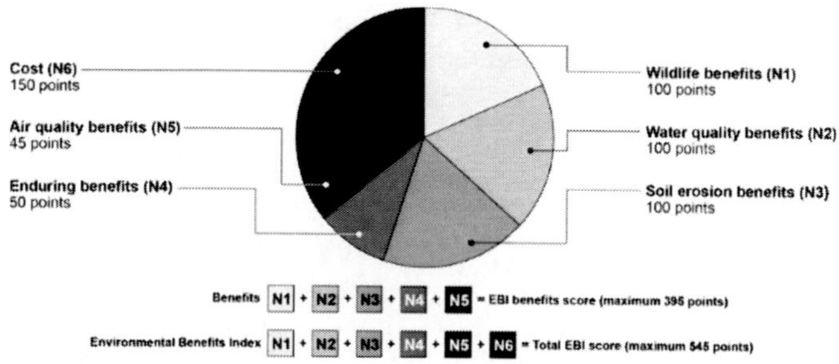

Source: U.S. Department of Agriculture (USDA). | GAO-24-106311.

USDA uses the letter N before each factor for general enrollment.

Wildlife benefits factor (N1): to encourage landowners and operators to plant cover on contract acreage that will be beneficial to wildlife.

Water quality benefits factor (N2): to evaluate the potential for improving water quality by reducing (1) sediment, nutrients, and other pollutants from entering water ways and aquifers; (2) downstream flood damage through the restoration of hydrology; and (3) the leaching of nitrates and pesticides into groundwater.

Soil erosion benefits factor (N3): to evaluate the on-farm benefits of reduced soil erosion.

Enduring benefits factor (N4): to evaluate the likelihood that the practice established will persist, be maintained, and provide benefits beyond the initial CRP contract period.

Air quality benefits factor (N5): to evaluate the air quality improvements from reducing airborne dust and particulate from cropland wind erosion that causes damage to nearby affected population concentrations. This factor also provides points for the value of land for carbon sequestration.

Cost factor (N6): to consider the relative cost of CRP offers by providing greater weight to offers with lower rental rates.

Figure 3. Maximum Points and Formula for the Environmental Benefits Index (EBI) Used to Score and Rank Offers for General Enrollment in USDA's Conservation Reserve Program (CRP), as of Fiscal Year 2023.

[25] For the cost factor in the EBI, USDA's formula for 125 out of 150 points is: a*(1-(Bid/b)). The bid represents a landowner's or operator's requested rental payment. The "a" represents the relative weight of cost factor as compared with environmental benefits score, and has been 125 since 1997, when it was changed from 190. The "b" is the highest rental payment in each signup and has been 240 since 2020 after having increased four times since 1997.

Some of the factors have subfactors. For example, the water quality benefits factor has a subfactor for location in a water quality zone identified by a state committee and submitted to USDA for approval.[26] The EBI assigns points to each factor and subfactor (as shown in fig. 3).[27] Points for soil erosion, water quality, and air quality are at least partially derived using science-based information from computer models. Points for other benefits are based on criteria found in the FSA's CRP handbook. These criteria are based on FSA, ERS, and NRCS officials' professional judgment. According to FSA officials, models are not yet available for all the factors and subfactors.

After assigning points for each offer, FSA ranks the offers by total points. For more information about the EBI, see appendix IX.

What Factors Does FSA Use to Select Offers for Grassland Enrollment?

To select offers for grassland enrollment, FSA uses an index as part of a competitive method to score offers. From 2020 through 2023, the grassland index has had seven ranking factors that total to determine the score. Four of these factors relate to environmental benefits, one relates to the demographic of the landowner or operator, one relates to the size of the livestock operation, and the seventh relates to the cost per acre of the offer (see fig. 4).[28]

---

[26] Water quality zones are intended to be areas where proposed land may contribute to groundwater or surface water quality impairment and assist in meeting certain federal, state, or local water quality laws. The zones are identified by state committees who are primarily active farmers and ranchers nominated by members of Congress and selected by the Secretary of Agriculture. The state committee may consult with a state technical committee that includes agricultural producers, certain forest landowners, and others who represent a variety of disciplines in soil, water, wetlands, plant, and wildlife sciences, including USDA and state representatives. The state committees submit their zones to USDA for approval.

[27] Criteria for indexes are in FSA's handbook for the program. See U.S. Department of Agriculture, *Agricultural Resource Conservation Program*.

[28] The beginning, limited resource, socially disadvantaged, or veteran farmers and ranchers factor is intended to encourage producers who meet certain demographic requirements to submit offers for their land. These requirements are determined by USDA, and USDA's form CCC-860 provides the definitions for each group. Landowners and operators complete this form to certify their inclusion in one or more of these groups. In addition, the small-scale livestock operation factor assigns points if the landowner certifies that they are eligible for a small-scale livestock operation, which is an operation with 140 grazing animal units or fewer. Different kinds of animals are assigned different units, per FSA's handbook for CRP. For example, a dairy cow is 1.40, a heifer is 0.86, and a goat is 0.15. Landowners and operators also certify that they will not offer more than 200 acres for the program.

The environmental factors include whether grassland preservation is maximized and whether threatened or endangered species are known to be in the area. The cost factor has two subfactors that assign points for offers (1) in counties in which the payment rate is less than or equal to $15 per acre or (2) that are below the corresponding county rental rate. FSA does not use science-based information from computer models to assign points to the factors and subfactors; rather, the points are based on a list of criteria in the CRP handbook, which was developed based on FSA officials' professional judgment with advice from other agencies. For certain subfactors, state committees identify which areas of their state should receive certain points and submit these areas to USDA for approval.[29] For more information about the grassland index, see appendix X.

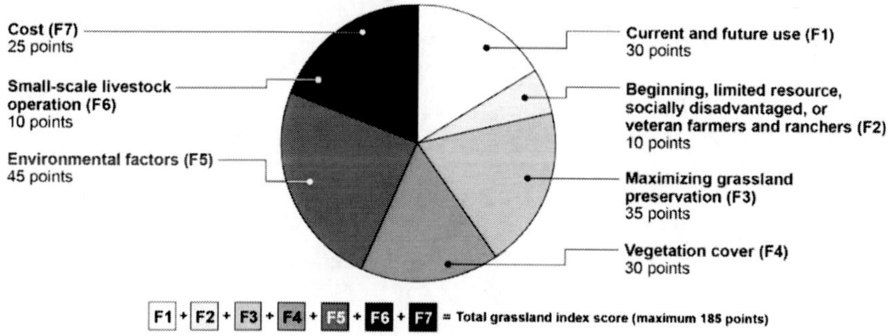

Source: U.S. Department of Agriculture (USDA). | GAO-24-106311.

Figure 4. Maximum Points and Formula Used to Score and Rank Offers for Grassland Enrollment in USDA's Conservation Reserve Program (CRP), as of Fiscal Year 2023.

How Does FSA Choose Offers for Continuous Enrollment?

For continuous enrollment, FSA automatically accepts offers of environmentally sensitive land devoted to certain conservation practices if the

[29] Certain zones are identified by state committees who are primarily active farmers and ranchers nominated by members of Congress and selected by the Secretary of Agriculture. The state committee may consult with a state technical committee that includes agricultural producers, certain forest landowners, and other professionals who represent a variety of disciplines in soil, water, wetlands, plant, and wildlife sciences, including USDA and state representatives. The state committees submit their zones to USDA for approval.

land and landowner or operator meet certain eligibility requirements and enrollment levels do not exceed the statutory cap.[30]

USDA's State Acres for Wildlife Enhancement (SAFE) Project on Migratory Birds, Butterflies, and Pollinators

The U.S. Department of Agriculture (USDA) established a project within its SAFE initiative within its Conservation Reserve Program (CRP) in 2017 in cooperation with partners. The project sought to enroll certain wetlands that provide critical habitat for migratory birds and other wildlife species.

These wetlands specifically include areas with playa lakes, which are round hollows in the ground that fill with water after spring rainstorms or are fed by water from underlying aquifers. They support a variety of wildlife, including waterfowl, and are located in states such as Colorado, Kansas, Nebraska, New Mexico, Oklahoma, and Texas.

An official from USDA's Farm Service Agency stated that before the SAFE project, USDA had difficulty enrolling these wetlands in CRP. The official said that farmers were reluctant to enroll these wetlands because they can be agriculturally productive during periods when the land is not wet. Thus, farmers were more inclined to keep the land in production than offer it for CRP.

The project resulted in increased enrollment of playa wetlands in parts of Kansas, Nebraska, and Oklahoma, according to USDA. As of August 2023, USDA had enrolled over 17,000 wetlands acres across Kansas, Nebraska, and Oklahoma through the project.

Playa Lakes

Sources: GAO analysis of USDA information; Natural Resources Conservation Service (photo). | GAO-24-106311.

---

[30] See 7 C.F.R. § 1410.31(d).

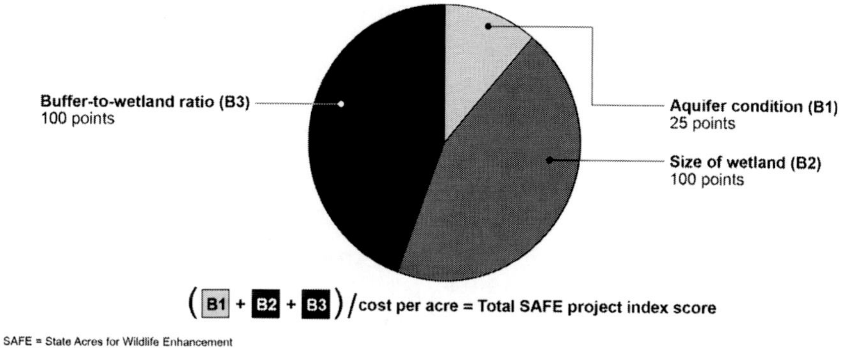

Source: GAO analysis of U.S. Department of Agriculture (USDA) information. | GAO-24-106311.

USDA uses the letter B before each factor for this project's index.

Aquifer condition factor (B1): to encourage enrollment of land situated over declining portions of an aquifer.

Size of wetland factor (B2): to encourage enrollment of larger wetlands. Wetlands are low-lying ecosystems where the water table is always at or near the surface. They perform vital ecological functions, including providing critical habitat for wildlife and waterfowl, mitigating floods by slowing down and absorbing excess water during storms, and purifying water by filtering out pollutants before they enter streams and lakes.

Buffer-to-wetland ratio (B3): to encourage enrollment of land that contains less buffer and more wetland. Buffers are borders of grass or trees or both along rivers, streams, or other waterways.

Figure 5. Maximum Points and Formula for Scoring and Ranking Offers for the SAFE Project on Migratory Birds, Butterflies, and Pollinators within Continuous Enrollment, as of Fiscal Year 2023.

Because FSA typically does not use a competitive method for continuous enrollment, it does not use specific scoring and ranking factors. Instead, FSA accepts offers throughout the fiscal year, except for one specific project.[31] Specifically, for four sign-ups from 2017 to 2023, FSA tested the use of competitive bidding with an index, rather than automatically accepting offers that met certain criteria, for a project within its State Acres for Wildlife Enhancement (SAFE) initiative under continuous enrollment. For two of those

[31] In addition to this project, another effort within CRP scores and ranks offers. Specifically, in FSA's CREP partnership with the state of Minnesota, the state is responsible for evaluating, scoring, and ranking offers as part of the application process. For land to be eligible for its CREP, Minnesota requires that it also be enrolled in the state's conservation easement program, and the state evaluates, scores, and ranks the offers for this program.

sign-ups (in 2022 and 2023), FSA sought to cost- effectively enroll high-benefit wetland acres through the SAFE project (see sidebar) by using a benefit-cost ratio—a ratio of benefit score divided by rental rate—to score and select offers rather than automatically enrolling acres that met certain criteria (see fig. 5).[32]

What Has Been the Compositionof Acres and Rental Payments in CRP?

The number of acres enrolled in CRP and amount of total rental payments have fluctuated over time. Total acreage generally increased from 1.9 million acres in FY 1986—the program's first year of enrollments—to a high of 36.8 million acres in FY 2007. It then decreased to 22.9 million acres in FY 2023, in part due to lower acreage maximums set by Congress.[33] In addition, the portion of land in continuous and grassland enrollments have increased since each type was introduced, while acres in general enrollment have decreased. Figure 6 shows acreage trends by type of enrollment since CRP began enrollments in 1986.

As figure 6 shows, for FY 2023, general enrollment and continuous enrollment each comprised more than a third of acres (8.4 million and 8.2 million, respectively), while grassland enrollment accounted for more than a quarter (6.3 million acres). Also, for FY 2023, general enrollment comprised

---

[32] The benefit-cost ratio scored and selected offers based on three factors that related to the condition of the aquifer, the size of the wetland, and the buffer-to-wetland ratio. Additionally, USDA ranked offers separately from seven focal areas in Kansas, Nebraska, and Oklahoma, within the SAFE eligibility area. USDA also determined that up to 50 percent of offers could be accepted in any of the sign-ups. Wetlands are low-lying ecosystems where the water table is always at or near the surface. They perform vital ecological functions, including providing critical habitat for wildlife and waterfowl, mitigating floods by slowing down and absorbing excess water during storms, purifying water by filtering out pollutants before they enter streams and lakes, among other things. See GAO, *Farm Programs: USDA Should Take Additional Steps to Ensure Compliance with Wetland Conservation Provisions*, GAO-21-241 (Washington, D.C.: Apr. 2, 2021).

[33] The following are the maximum CRP acreage set by each farm bill: 1985 Farm Bill, 45 million acres; 1990 Farm Bill, 45 million acres; 1996 Farm Bill, 36.4 million acres; 2002 Farm Bill, 39.2 million acres; 2008 Farm Bill, 32 million acres; and 2014 Farm Bill, 27.5 million acres in FY 2014, 26 million acres in FY 2015, 25 million acres in FY 2016, and 24 million acres in FY 2017 and FY 2018. Beginning with the 2018 Farm Bill, the acreage limitation gradually increased until it reached 27 million acres in FY 2023. The 2018 Farm Bill established maximum and minimum levels per year for acres enrolled in CRP. In fiscal year 2023, the maximum for the overall CRP was 27 million acres, the minimum was 8 million acres for continuous enrollment, and 2 million acres for grassland enrollment.

about a quarter of rental payments ($480 million); continuous enrollment, two-thirds ($1.2 billion); and grassland enrollment, about 5 percent ($99 million).

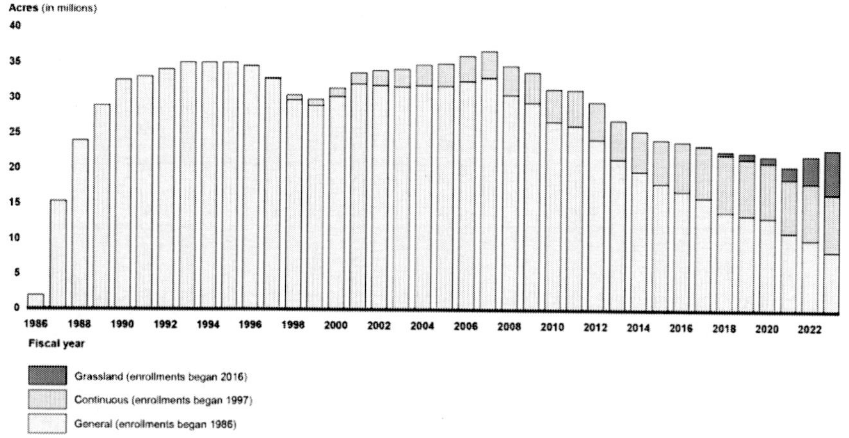

Source: GAO analysis of U.S. Department of Agriculture (USDA) information. | GAO-24-106311.

Figure 6. Acres Enrolled in USDA's Conservation Reserve Program, by Enrollment Type, Fiscal Years 1986–2023.

Total annual rental payments ranged between $1.3 billion and $1.8 billion (not adjusted for inflation) per year from FY 1989 through FY 2023. As with acres enrolled, the portion of total rental payments for continuous and grassland enrollment have generally increased since their introduction, while the portion for general enrollment has generally decreased since 2006. In addition, total annual rental payments for continuous enrollment have exceeded those for general enrollment since fiscal year 2016. Figure 7 shows total annual rental payments by type of enrollment since CRP's implementation.[34]

[34] During our review, we identified contracts for which rental rates were higher than USDA's maximum rate. On March 28, 2024, we notified FSA of this deficiency. Based on the preliminary results from FSA's investigation, FSA officials stated that they estimated the overpayments to be about $180,000 for 126 contracts from FY 2017 through FY 2023. This represents less than 0.01 percent of the $12.5 billion in rental payments that USDA made over this time. The FSA officials said that this error likely occurred when the maximum rental rates were uploaded into new software prior to the FY 2017 contract payments. As of April 2024, USDA had not completed its investigation. We plan to follow up with USDA on the results of its investigation and any corrective actions based on this investigation.

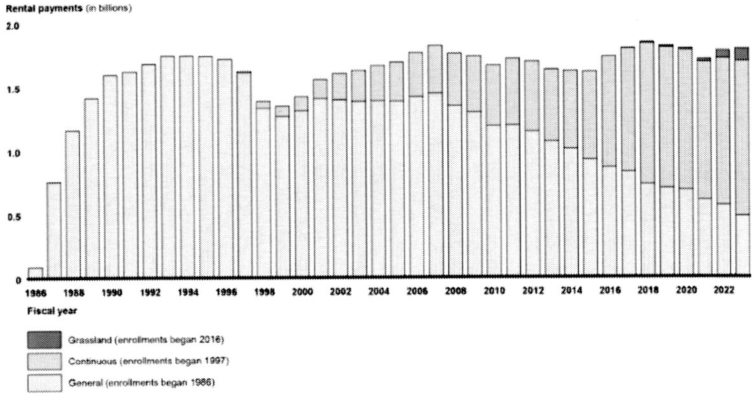

Source: GAO analysis of U.S. Department of Agriculture (USDA) information. | GAO-24-106311.

Figure 7. Total Rental Payments (not adjusted for inflation), by Type of Enrollment, for USDA's Conservation Reserve Program, Fiscal Years 1986–2023.

When adjusted for inflation, total CRP rental payments peaked in the early 1990s, decreased until FY 1999, and then stayed within a narrow range until FY 2007, when total inflation-adjusted payments were about $2.5 billion. Since then, total inflation-adjusted payments have generally decreased (see fig. 8), reaching $1.8 billion in FY 2023. This decrease generally coincided with the decrease in total CRP acres.

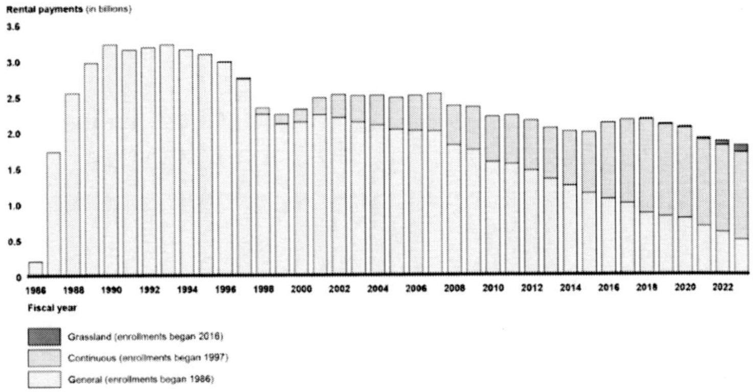

Source: GAO analysis of U.S. Department of Agriculture (USDA) information. | GAO-24-106311.

Note: We adjusted the rental payments for inflation to fiscal year 2023 dollars.

Figure 8. Total Rental Payments (adjusted for inflation), by Type of Enrollment, for USDA's Conservation Reserve Program, Fiscal Years 1986–2023.

Appendixes I and II provide more information on historical CRP rental payments per acre and how long acres have been enrolled in the program.

What Was the Geographic Distribution of Acres and Rental Payments for FY 2023?

The distribution of acres and payments varied geographically. About 57 percent (13 million acres) of acres enrolled in FY 2023 were in seven states in the central U.S., and total rental payments varied across states but did not necessarily correlate with acreage. Figure 9 shows acreage enrolled and rental payments by state for fiscal year 2023.

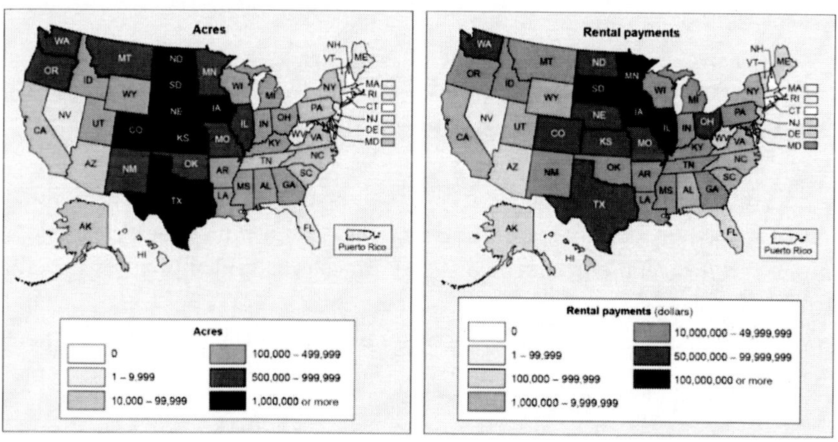

Source: GAO analysis of U.S. Department of Agriculture (USDA) information. Map Resources(map). | GAO-24-106311.

Figure 9. Total Acres Enrolled and Rental Payments, by State or Territory, in USDA's Conservation Reserve Program, Fiscal Year 2023.

The seven states comprising 57 percent of acres enrolled—Colorado, Iowa, Kansas, North Dakota, Nebraska, South Dakota, and Texas— enrolled at least 1 million acres each. The states with the most acres in general enrollment were the Great Plains states of Colorado, Kansas, and Texas.[35] Continuous enrollment was most concentrated in the Dakotas, Illinois, Iowa,

[35] The Great Plains in the U.S. include at least portions of 10 states: Colorado, Kansas, Montana, New Mexico, Nebraska, North Dakota, Oklahoma, South Dakota, Texas, and Wyoming.

and Minnesota. Grassland enrollment was most concentrated in the Great Plains states of Colorado, Nebraska, New Mexico, and South Dakota.

Almost half (48 percent) of total rental payments went to four states—Illinois, Iowa, Minnesota, and South Dakota—which received at least $100 million each. The states with the highest rental payments varied by type of enrollment. For example, the five states that received the most in rental payments for continuous enrollment were Illinois, Iowa, Minnesota, Missouri, and South Dakota. In contrast, the five states that received the most in rental payments for grassland enrollment were Colorado, Kansas, Nebraska, New Mexico, and South Dakota. For more information about the distribution of acres and rental payments by state in FY 2023, see appendix III.

We also analyzed county-level distribution of acres and rental payments; appendixes IV and V present the results of these analyses.

How Do Acreage Per Contract and Acreage Per Conservation Practice Vary By Enrollment Type For FY 2023?

For FY 2023, we found that acreage per contract and acreage per conservation practice varied by type of enrollment. For example, median acreage per contract for grassland enrollment was five times greater than that for general enrollment. Also, grass and legume practices represented 71 percent of general enrollment and 28 percent of continuous enrollment. Table 2 provides more information on these topics. For more information on acres and rental payments by conservation practices, see appendix VI.

Table 2. Acreage and conservation practices by type of enrollment in USDA's Conservation Reserve Program (CRP), fiscal year 2023

	General	Continuous	Grassland
Median number of acres per contract	30 acres	6 acres	151 acres
Range in acres per contract[a]	1 to 603 acres	0.3 to 189 acres	4 to 3,326 acres
Largest conservation practices (percent of acres in type of enrollment)[b]	Establishment or maintenance of: Permanent native grasses (45%) and introduced grasses and legumes (26%).	Establishment or maintenance of: Permanent native grasses (14%), introduced grasses and legumes (14%), permanent wildlife	Maintenance of: Permanent grass and legumes on grassland[f] (98%).

	General	Continuous	Grassland
		habitat (10%), filter strips[c] (8%), riparian buffers[d] (6%), and duck nesting habitat (6%). Also, restoration of wetlands[e] (non-floodplain) (10%).	

Source: GAO analysis of U.S. Department of Agriculture (USDA) information. | GAO-24-106311.

Note: Legumes are plants that have seeds in a pod and improve soil by taking nitrogen from the air.

[a] To provide a typical range for CRP contract acreage without reference to extreme outliers, we excluded the CRP contracts with the lowest and highest 1 percent of entries.

[b] We selected the largest conservation practices that sum to at least two-thirds of the acreage for each type of enrollment.

[c] A filter strip is a narrow band of grasses, legumes, and forbs—plants that are not woody and have broad leaves. Filter strips improve water quality by intercepting sediment and nutrients, mitigate erosion by reducing the negative impacts of wind and water, and provide habitat and corridors for wildlife.

[d] Riparian buffers are strips of trees bordering streams, wetland areas, and other waterbodies to improve water quality and wildlife habitat.

[e] Wetlands are low-lying ecosystems where the water table is always at or near the surface. They perform vital ecological functions, including providing critical habitat for wildlife and waterfowl, mitigating floods by slowing down and absorbing excess water during storms, purifying water by filtering out pollutants before they enter streams and lakes, among other things.

[f] This practice applies only to grassland enrollment and the land may be used for haying or grazing.

USDA May Have Opportunities to Further Improve How It Selects Offers for CRP but Conducts Limited Reviews of Its Selection Methods

According to 17 studies and eight experts, USDA has opportunities to further improve its methods for selecting CRP offers to increase the program's overall benefits and cost-effectiveness. FSA has taken steps to use science-based information, through updated computer models, to assign points for scoring certain environmental benefits. FSA uses this information to more accurately estimate the environmental benefits for CRP offers, according to FSA officials. However, the agency has not begun using science-based information to score other benefits, according to agency officials. Additionally, FSA does not have a process for periodically reviewing its selection methods, which includes soliciting and publicly reporting experts' suggestions on these methods, and updating the methods as needed.

What Benefits Do FSA's Current Selection Methods Have, According to Studies and Experts?

FSA's current selection methods have various benefits, including potentially increasing environmental benefits, according to two studies we reviewed and eight experts we interviewed.[36] For example, the introduction of the EBI expanded the environmental benefits that CRP targets beyond soil erosion to include improvements in wildlife habitat, water quality, and regional conservation priorities, according to two studies.[37] Furthermore, according to one of those studies, using the EBI, rather than the prior methods that were based on soil erosion alone, to select offers might have increased CRP benefits associated with freshwater recreation, wildlife viewing, and pheasant hunting from $459 million to $828 million per year.[3838]

Additional benefits of the EBI and other indexes include their simplicity, transparency, and cost-effectiveness, according to experts, who include academics and current and former FSA and ERS officials:[39]

- The indexes are fairly easy for FSA to calculate using available data.
- The indexes provide transparency because landowners and operators as well as researchers can see how the scores are calculated.
- The indexes are useful in selecting offers. In addition, the EBI increased the program's cost-effectiveness, according to an ERS study and two experts.

---

[36] U.S. Department of Agriculture, Economic Research Service, Conservation Program Design: Better Targeting, Better Outcomes, Economic Brief No. 2 (March 2006). The 2006 paper provided the estimates which were based on work from the following paper: U.S. Department of Agriculture, Economic Research Service, Economic Valuation of Environmental Benefits and the Targeting of Conservation Programs: The Case of the CRP, Agricultural Economic Report No. 778 (Washington, D.C.: April 1999).

[37] An EBI was first developed for use in ranking offers in 1991, but ranking did not play a prominent role in determining offer acceptability until 1997. John Johnson et al., *Stakeholders Look at the Conservation Reserve Program Through Different Windows*, The Conservation Reserve Program–Planting for the Future: Proceedings of a National Conference (Fort Collins, CO: June 2004).

[38] U.S. Department of Agriculture, Economic Research Service, *Conservation Program Design: Better Targeting, Better Outcomes*.

[39] The experts were generally less familiar with the indexes used for the grassland sign-up and for the SAFE project on migratory birds and other wildlife within continuous sign-up. They based their comments on these indexes on copies of the indexes that we provided for their review prior to interviews.

However, eight experts identified opportunities to increase the program's cost-effectiveness, as described below.

To what extent Does FSA Use Science-Based Information to Help Select CRP Offers?

FSA uses science-based information from computer models to help assign points for some of the benefits in the EBI. These points are used to help select CRP offers for general enrollment, but FSA is not using such information for the benefits in the grassland index. In addition, FSA has completed some efforts, and has others under way, to increase the use of science-based information for assigning points in these indexes. FSA officials stated that these efforts can provide additional objective, site-specific information to further improve land selection, resulting in increased environmental benefits. When science-based information from computer models is not available, FSA relies on a criterion found in FSA's CRP handbook, which was developed based on the professional judgment of USDA officials with input from state committees and their advisors who identify zones that can receive points specified in the EBI.[40]

In the EBI, FSA uses science-based information from computer models to derive at least some of the points for soil erosion benefits, water quality benefits, and air quality benefits, but none of the points for wildlife benefits or enduring benefits.[41] In 2023, FSA began using updated models to more accurately estimate rates of soil erosion and associated surface runoff for land offered for CRP, according to an FSA document and FSA officials. One model incorporates site-specific biophysical processes—such as rainfall, rate of runoff, soil properties, and slope—to provide more reliable estimates for average annual soil loss related to rainfall. The other model simulates the effect of wind erosion, and both help with assigning points for soil erosion, water quality, and air quality. In addition, the improved accuracy of the

[40] USDA's point system assigns points, for example, based on the number and types of species of plants and whether they are native or introduced. It also assigns points based on whether the area is in a zone (such as for wildlife, air quality or water quality) identified by a state committee with advice from a state technical committee and approved by USDA.

[41] Enduring benefits are the likelihood that the practice will provide benefits beyond the contract period. USDA does not consider soil erosion or water quality in its grassland index because these are not the purposes of grassland enrollment. Rather, the program emphasizes support for grazing operations, plant and animal biodiversity, and grassland and land containing shrubs and forbs under the greatest threat of conversion, according to USDA.

updated models may enhance CRP's cost-effectiveness because the EBI now gives more points to land with the potential for greater soil erosion, according to FSA officials.

However, FSA did not adopt these models for CRP until 13 to about 20 years after they were developed and other USDA agencies began using them. FSA officials said that incorporating the updates into USDA's older computer systems took over 3 years.[42]

In addition, FSA has taken steps to increase the use of science-based information, based on models, to assign points for wildlife and other benefits in the EBI and grassland index. For example, in 2020, FSA contracted with the Department of the Interior's U.S. Fish and Wildlife Service (FWS) to use hundreds of broad-scale bird density models to identify areas that host a diverse selection of bird species, according to FWS officials. Species were placed in groups that have similar requirements and are associated with grassland or forests. This was done so optimal CRP locations could be identified to provide wildlife benefits that align with the conservation practice being delivered (e.g., grass restoration versus woody planting).[43] FWS completed its update of the wildlife-related science-based information in March 2024, according to FSA and FWS officials. However, it is unclear whether FSA will incorporate the science-based information from the updated model into its indexes. We asked, but FSA officials had not yet decided whether they planned to use it to help score and select offers for general or grassland enrollments. FSA currently provides points for wildlife locations based on generally county-level zones identified by state committees and their advisors.

In another effort to increase the use of science-based information, as part of the same 2020 FSA contract, FWS developed a model to improve the accuracy of information used to assign points for one part of the grassland index—the risk that grassland will be converted to cropland or development. FSA currently estimates this risk using various factors, such as the percentage of grassland in that county. The new model, which FWS completed in 2023,

---

[42] In addition, FSA officials had concerns that landowners would object to the new model if their offers were rejected based on a low EBI score. Thus, USDA notified landowners whose offers were rejected on this basis that they could request USDA re-evaluate their offer. However, few bidders made this request, according to FSA officials.

[43] Birds are used as a bioindicator of ecosystem health affecting wildlife because (1) their ranges and environment associations are well understood; (2) they use all food and nutrition levels of the ecological pyramid—that is, birds eat plant seeds, insects, frogs, fish, and small mammals; (3) they have demonstrated response to environmental conditions; and (4) they are easily identified and measured.

predicts the risk of grassland conversion at a local level over the next 20 years. The model uses historical data from the past 20 years on grassland loss and its relationship to land cover, soil property, topography, climate, and socioeconomic factors, according to FSA and FWS officials.[44] FSA officials stated they had not decided whether they planned to incorporate the science-based information from this model into their grassland index, as of February 2024.

Historically, FSA has not updated its indexes to incorporate the use of science-based information from available computer models in a timely manner. USDA's core values include basing "decisions and policy on science and data that are reliable, timely, relevant, and free from political interference," according to USDA's *FY 2023 Performance Plan and Fiscal Years 2022–2026 Strategic Plan*.[45] In addition, our key practices for evidence-based policymaking include using evidence to learn and applying learning to decision-making, such as using evidence to inform management decisions.[46] By building on its existing efforts by incorporating, in a timely manner, reliable science-based information from models or other sources into the agency's methods for selecting offers, as appropriate, FSA may be able to more accurately estimate the benefits of CRP offers, increasing the program's environmental benefits and cost- effectiveness.

What Opportunities Exist for FSA to Improve Its Selection Methods, According to Studies and Experts?

According to 17 studies we reviewed and eight experts we interviewed, FSA has opportunities to further improve its methods for selecting CRP offers to further increase the program's benefits and cost-effectiveness, consistent with CRP's objectives.[47] These include the following:

---

[44] A manuscript describing this work has been accepted for publication in the Journal of Conservation Biology, according to FWS officials from the Habitat and Population Evaluation Team.

[45] U.S. Department of Agriculture, FY 2023 Performance Plan (Washington, D.C.: March 2022); and Strategic Plan: Fiscal Years 2022–2026 (Washington, D.C.: March 2022).

[46] GAO-23-105460.

[47] Specifically, CRP's objectives are "to cost-effectively reduce water and wind erosion, protect the Nation's long-term capability to produce food and fiber, reduce sedimentation, improve water quality, create and enhance wildlife habitat, and other objectives including, as appropriate, addressing issues raised by State, regional, and national conservation initiatives and encouraging more permanent conservation practices, including, but not limited to, tree planting." 7 C.F.R. §1410.3(c).

- *Use benefit-cost ratio.* FSA could adjust the indexes to use benefit-cost ratios—benefit points divided by rental rate—so that they measure cost-effectiveness, according to five experts and two studies. For example, if two offers scored 300 points for benefits and their rental rates were $100 and $50, respectively, the first offer would receive a score of 3 (points per dollar) and the second a score of 6 (points per dollar). In contrast, using EBI, the two offers would receive 73 points and 99 points, respectively. Thus, the benefit-cost ratios are proportional to costs, but EBI points are not. Using a benefit-cost ratio could have substantially increased acreage and environmental benefits for the same cost when compared with using the EBI, according to a 2016 study.[48] FSA officials stated that if they were to use this methodology for enrollments, the agency would have to consider statutory language on state historic enrollment requirements.[49]
- *Increase reliance on science-based information, such as from models.* Increasing reliance on science-based information, such as from models or other sources, rather than professional judgment to assign points could ensure the indexes better reflect actual benefits, according to three experts. For example, one expert suggested developing science-based information that could be converted to assign points for climate change benefits within the EBI.
- *Adjust points assigned to certain benefits.* For example, increasing the points for establishing native plants—particularly native wildflowers—that support insects and pollinators could produce better outcomes, according to three experts. Such practices have a high level of benefits but can be substantially more expensive for landowners and operators to plant, as compared with basic grass cover, according to a 2022 ERS study.[50] The study states that as a

---

[48] Ruiqing Miao et al., "Assessing Cost-effectiveness of the Conservation Reserve Program (CRP) and Interactions between the CRP and Crop Insurance," *Land Economics*, vol. 92, no. 4: 593–617.

[49] State historic enrollment requirements refers to a provision in the 2018 Farm Bill. The 2018 Farm Bill states that at the beginning of each of fiscal years 2019 through 2023, to the maximum extent practicable, the Secretary of Agriculture shall allocate to the states proportionately 60 percent of the available number of acres each year for enrollment in the conservation reserve, in accordance with historical state enrollment rates.

[50] U.S. Department of Agriculture, Economic Research Service, *Cover Practice Definitions and Incentives in the Conservation Reserve Program*, Economic Information Bulletin 233 (Washington, D.C.: February 2022).

result, landowners and operators may lack an incentive to establish such beneficial plants.

- *Consider alternatives to bid caps.* To overcome difficulties with setting bid caps (maximum allowable rental rates), FSA could consider using a different approach, according to five experts. Bid caps tend to be low, according to five experts. Bid caps that are too low can discourage landowners and operators from submitting offers, reducing competition among the remaining, potentially more expensive offers, according to three studies.[51] The remaining offers face little competition to reduce their bids to increase their likelihood of enrolling into CRP. This makes the remaining offers more expensive than when they face higher competition to enroll in the program. These studies suggested FSA use an approach that allows greater flexibility for participants to bid what they are willing to accept to enroll in the program. For example, FSA could use information that it maintains or the bidding process itself to identify appropriate rental payment rates, according to a 2021 study.[52]
- *Expand use of competition for continuous enrollment.* Increasing the use of a competitive process would enable FSA to measure the benefits and cost-effectiveness of offers for continuous enrollment, according to two experts. These experts said that the competitive process could be especially useful in acquiring the most cost-effective offers for similar practices, such as grass practices, that general and continuous enrollment have in common.

Seven experts identified challenges FSA could face in pursuing some of these opportunities, including political, stakeholder, and information technology challenges. For example, some said changes in methods— such as adjusting an index to use a benefit-cost ratio—could shift the location of CRP land and rental payments from certain counties and states to others. FSA

---

[51] U.S. Department of Agriculture, Economic Research Service, *Options for Improving Conservation Programs: Insights From Auction Theory and Economic Experiments*, Economic Research Report No. 181 (Washington, D.C.: January 2015); Daniel Hellerstein et al., "The Effective Use of Limited Information: Do Bid Maximums Reduce Procurement Cost in Asymmetric Auctions?," *Agricultural and Resource Economics Review*, vol. 39, no. 2 (2010): 288–304; and Peter Cramton et al., "Improving the Cost-Effectiveness of the Conservation Reserve Program: A Laboratory Study," *Journal of Environmental Economics and Management*, vol. 108 (2021).

[52] Peter Cramton et al., "Improving the Cost-Effectiveness of the Conservation Reserve Program: A Laboratory Study."

officials agreed that stakeholders might object to certain changes in methods because these changes could affect their land, payments, or environmental benefits. These officials also said that USDA's software and hardware are dated, making some changes difficult.[53]

To What Extent Does FSA Review Its Selection Methods to Determine Whether They Need Updating?

While FSA officials have periodically discussed the EBI and grassland index in meetings, FSA has conducted limited reviews of its selection methods (i.e., how it selects CRP offers) over the past 20 years or so, and generally did not have much documentation of these review meetings, according to these officials. FSA officials were able to provide limited records related to a few meetings, such as the agenda for a meeting held in 2009 with ERS and other agencies, and they stated that the most recent meeting with staff from other agencies to discuss EBI was in 2019.[54]

Over the same time frame, FSA made a few, generally small, updates to its methods—specifically, the EBI—for selecting CRP offers. For example, FSA periodically adjusted the two parameters it uses to calculate the cost factor—one related to the maximum bid cap, and the other related to the weight of the cost factor relative to the environmental benefits score. FSA also made a few small changes related to determining points for the EBI, according to a summary officials provided. For example, FSA eliminated the points it previously assigned to the wildlife habitat monoculture cover—a cover composed of a single plant species.[55] In addition, as described above, in 2023 FSA began using data from two updated computer models to assign points for soil erosion and water quality benefits in the EBI.

Although former FSA and ERS officials stated that FSA's intention when it developed the EBI was to change it as research became available to support

---

[53] In addition, ERS officials told us they were uncertain whether USDA had the authority to implement the suggestion to use an alternative method to bid caps.

[54] In addition, FSA officials meet after each general sign-up to discuss how well the sign- up worked. Also, in 2009, USDA held a meeting on the EBI with officials from the Environmental Protection Agency, Department of Commerce (National Oceanic and Atmospheric Administration), and Department of the Interior FWS and the U.S. Geological Survey).

[55] The points the EBI previously assigned to wildlife habitat monoculture cover comprised 10 points of the maximum 395 total EBI benefit points.

improvements, key aspects of the EBI have changed little over the past 20 years. Key EBI characteristics that have not changed include the following:

- The EBI method of adding a benefit score to a cost factor has not changed since 1997.
- The environmental benefits categories (e.g., air quality) and the maximum points assigned to each category have not changed since 2003.

Two former FSA and ERS officials said the agency has been slow to implement changes that would enhance CRP due to concerns over the potential ramifications of upsetting stakeholders—including congressional, farm, and environmental stakeholders—if acres and payments are redistributed to other counties or states. They also cited concerns over difficulties in implementing changes with outdated information systems. One former official cited the 20-year delay in updating the soil erosion model as an example. Current FSA officials agreed that there may be opportunities to improve the selection methods and that challenges may arise from stakeholders and outdated information systems.

A 2001 study noted that the EBI was not intended to be a rigid index, but to be adjusted and improved depending on the progress of sign-ups, perceived deficiencies, and changed priorities.[56] Furthermore, USDA's FY 2023 Performance Plan and Fiscal Years 2022–2026 Strategic Plan cite USDA's core value of basing "decisions and policy on science and data that are reliable, timely, relevant and free from political interference."[57]

However, FSA has only conducted limited reviews of its methods for selecting CRP offers because it does not have a process for doing so or for routinely soliciting expert views on these methods, according to FSA officials. USDA's Fiscal Years 2022–2026 Strategic Plan places an emphasis on building evidence to inform decision-making and calls for the agency to conduct an analysis of alternative decision-making strategies for USDA's conservation programs, including CRP.[58] In addition, our key practices for

---

[56] Marc O. Ribaudo et al., "Environmental Indices and The Politics of the Conservation Reserve Program," *Ecological Indicators*, vol. 1, no. 1 (2001): 11-20.

[57] U.S. Department of Agriculture, FY 2023 Performance Plan and Strategic Plan: Fiscal Years 2022–2026.

[58] Specifically, USDA stated this to achieve objective 1.3, "Restore, Protect, and Conserve Watersheds to Ensure Clean, Abundant, and Continuous Provision of Water Resources." See U.S. Department of Agriculture, *Strategic Plan: Fiscal Years 2022–2026.*

evidence-based policymaking include (1) assessing the evidence, such as establishing routine processes to identify and assess relevant evidence; and (2) involving a range of stakeholders, such as researchers, which we have found is often vital to the success of federal efforts.[59] In addition, these key practices include using evidence to learn and apply learning to decision-making, such as using evidence to inform management decisions. In this case, FSA could address findings resulting from reviews of its CRP selection methods as a way to use evidence and apply it in agency decision making. As described above, experts have identified potential opportunities for FSA to adjust its methods in ways that could further improve the program.

FSA officials generally agreed that developing a process for reviewing how FSA selects CRP offers could be helpful in increasing the program's cost-effectiveness. FSA officials said that including academics and others in an expert panel could also be beneficial. However, they said that the panel would need to understand the realities of managing the program and that some changes might take a long time to implement. In addition, all eight experts we interviewed generally agreed that a process for reviewing these methods should include steps for soliciting reliable and timely information from scientists and economists, including those not in USDA, and publicly reporting on suggestions they make for improvements to FSA's methods. GAO's key practices for evidence- based policymaking highlight the importance of communicating relevant information internally to leaders and staff and externally to key stakeholders, such as Congress.[60]

By developing a process to periodically review its methods for selecting CRP offers, FSA would be better positioned to identify potential changes, based on reliable and timely science-based information. Including in this process a step to solicit and publicly report on suggestions from experts would help the agency ensure key stakeholders, such as Congress and researchers, understand the steps FSA may take to better achieve CRP's objectives. Finally, by addressing any findings resulting from its periodic reviews, as appropriate, FSA may increase the environmental benefits and cost-effectiveness of CRP.

---

[59] GAO-23-105460.

[60] GAO-23-105460.

FSA Has Analyzed and Reported CRP's Environmental Benefits to Some Extent

FSA uses science-based information from computer models to estimate some key environmental benefits. However, it has not analyzed environmental benefits by type of enrollment and has not published annual reports on CRP's environmental benefits since 2017.

To What Extent Does FSA Use Science-Based Information to Estimate Key Environmental Benefits?

For some key environmental benefits, FSA has used science-based information from computer models to estimate annual soil erosion and water quality benefits from CRP since at least 2006. For example, FSA has a model incorporating science-based information on site-specific biophysical processes—such as rainfall, soil properties, and slope—to provide these estimates for average annual soil loss related to rainfall and related water quality. In 2023, FSA officials said they began using updated computer models to re-estimate these environmental benefits more accurately.

In addition, FSA has been working with FWS to incorporate the use of the science-based information on wildlife that FSA obtained from FWS to estimate overall wildlife benefits.[61] The new model can use data from USDA's information systems to estimate wildlife benefits for each CRP site, practice, and enrollment type, or for the entire program, according to documents on the effort as well as FSA and FWS officials. Having information on wildlife benefits by enrollment type and for CRP overall, for the first time, can provide insights to USDA managers and Congress on CRP's performance in this area and whether changes are needed.

---

[61] USDA has partnered with a variety of organizations to conduct studies on the benefits of CRP, including for select animal species and locations. For example, USDA contracted with the FWS for a study which estimated that between 1992 and 2012, CRP resulted in the hatching of over 37 million additional ducks from the North Dakota, South Dakota, and northeastern Montana portion of the prairie pothole region, which includes wetlands that serve as habitat.

To What Extent Has FSA Analyzed and Reported Environmental Benefits by Type of Enrollment?

FSA generally has not analyzed and reported environmental benefits by type of enrollment and has not reported on environmental benefits for CRP as a whole since FY 2017. Specifically:

- *Benefits by type of enrollment.* FSA generally has not analyzed and reported on environmental benefits for CRP by type of enrollment, with one exception. FSA has reported on water quality benefits for conservation practices used in continuous enrollment, most recently for FY 2021.[62] In June 2023, we discussed with FSA officials the feasibility of analyzing each environmental benefit separately for each type of enrollment, and they told us it was not feasible. However, in September 2023, they told us that they had identified a way to estimate environmental benefits by type of enrollment.
- *Benefits for the entire CRP.* FSA previously reported on benefits for the entire CRP for FY 2006 through FY 2017 on its website. However, an FSA official told us that FSA stopped reporting these benefits after FY 2017 because it was not a priority for the administration at that time, and the official who developed the reports had retired.

In May 2024, an FSA official told us they expect to analyze and report updated data on environmental benefits, including water quality benefits, for 2006 through 2023 by the end of the fiscal year. According to this official, they expect to do so for the entire program and by enrollment type. The FSA official said they also plan to report data on wildlife benefits for the entire program and by type of enrollment for FY 2023 by the end of FY 2024.

However, it is uncertain whether FSA will publicly report this information annually in the future, given prior lapses in reporting. USDA's annual performance plans cite USDA's core value of basing decisions and policy on science and data that are reliable, timely, and relevant. In addition, GAO's key practices for evidence-based policymaking highlight the importance of

[62] The 2018 Farm Bill requires USDA to annually report on estimated water quality benefit summaries, with respect to continuous water quality practices. Pub. L. No. 115-334, § 2201(a)(3), 132 Stat. 4490, 4533 (codified at 16 U.S.C. § 3831(d)(3)(D)).

communicating relevant information internally to leaders and staff, and externally to key stakeholders, such as Congress.[63]

By annually analyzing and publicly reporting on CRP's environmental benefits, including by enrollment type and the entire program, FSA and external stakeholders would better understand the distinct benefits of each enrollment type in CRP and their performance over time. Such information could enable legislators to make better informed decisions about each type of enrollment, such as in setting acreage minimums or caps. This information could also help stakeholders such as landowners, operators, and environmental organizations better understand the benefits of, and make informed decisions about, participating in the program.

Conclusion

CRP seeks to offset detrimental effects that agricultural practices can have on the environment by paying landowners and operators to conserve land using practices that achieve certain environmental benefits. The methods FSA uses to select offers of land for CRP have some benefits. Opportunities exist to increase the program's environmental benefits and cost-effectiveness, according to experts and studies by academia and USDA.

To select land for the program, FSA relies on science-based information to assign points for certain environmental benefits, but not for others, in its index for general enrollments. To assign points for other benefits for general enrollment, and for grassland enrollment, FSA relies on less objective methods that are based on FSA staff's professional judgment, with advice from other agencies. However, FSA has recently developed additional science-based information and computer models to, for example, provide estimates of wildlife benefits on different types of land. FSA has not yet decided whether to incorporate this information into its indexes. By incorporating science-based information—based on models or other sources that FSA determines to be appropriate and reliable—to assign points for benefits, FSA could more accurately estimate the environmental benefits of each offer for general and grassland enrollment. Doing so could increase the overall benefits that the program achieves for its cost.

Further opportunities exist to increase the program's environmental benefits and cost-effectiveness, according to experts and studies by USDA and

[63] GAO-23-105460.

academia. However, FSA has made generally small changes to its selection methods—specifically, the EBI—over the past 20 years. FSA does not have a process for periodically reviewing these methods that includes soliciting and publicly reporting experts' suggestions for improvements. Involving a range of stakeholders, such as researchers, is supported by key practices for evidence-based policymaking that we have highlighted in our prior work. By developing and implementing such a process, FSA could help ensure it identifies and capitalizes on opportunities to increase CRP's environmental benefits and cost- effectiveness.

FSA previously published annual summaries of the environmental benefits for the entire CRP but has not done so since FY 2017 and has not analyzed and published these benefits by enrollment type. Analyzing and reporting on environmental benefits by type of enrollment—general, continuous, and grassland—could help USDA managers, Congress, and other stakeholders better understand the different benefits of each enrollment type and make better informed decisions about the program.

Recommendations for Executive Action

We are making the following four recommendations to USDA:

The Administrator of USDA's FSA should build on existing efforts by incorporating, in a timely manner and as appropriate, reliable science- based information from models or other sources into the agency's methods for assigning points that are used to select offers for CRP. (Recommendation 1)

The Administrator of USDA's FSA should develop a process to periodically review its methods for selecting offers for CRP. Such a process should specify the frequency of the reviews and include soliciting and publicly reporting on suggestions from experts, including non-USDA scientists and economists. (Recommendation 2)

The Administrator of USDA's FSA should address any findings resulting from its periodic reviews to enhance its methods for selecting offers for CRP, as appropriate. (Recommendation 3)

The Administrator of USDA's FSA should annually analyze information on environmental benefits by type of enrollment and for the entire program, and publicly report the information, such as on its website. (Recommendation 4)

Agency Comments and Our Evaluation

We provided a draft of this report to USDA for review and comment. FSA provided written comments and ERS provided technical comments, which we incorporated as appropriate. In FSA's comments, reproduced in appendix XI, it generally agreed with all four recommendations but noted challenges with the first.

Specifically, while FSA agreed that using recent studies and models would provide a more reliable, science-based approach to assigning points for selecting CRP offers, it noted that implementing certain updates to the current scoring methods could require significant resources and an analysis of impacts to the program, and could take considerable time. It also stated that any new legislation, such as the farm bill, could affect FSA's implementation of this recommendation. As discussed earlier in the report, we acknowledge the challenges that FSA may experience in updating its current scoring methods. We maintain that the effort would be beneficial because FSA could then more accurately estimate the benefits of CRP offers, increasing the program's environmental benefits and cost- effectiveness.

We are sending copies of this report to the appropriate congressional committees, the Secretary of Agriculture, and other interested parties. In addition, the report is available at no charge on the GAO website at https://www.gao.gov.

If you or your staff have any questions about this report, please contact me at (202) 512-3841 or morriss@gao.gov. Contact points for our Offices of Congressional Relations and Public Affairs may be found on the last page of this report. GAO staff who made key contributions to this report are listed in appendix XII.

Steve D. Morris

Steve D. Morris
Director, Natural Resources and Environment

Appendix I: Information on Average Rental Payments per Acre

Since the establishment of the Conservation Reserve Program (CRP), administered by the U.S. Department of Agriculture's (USDA) Farm Service Agency, the average rental payment per acre for the program as a whole and by type of enrollment has generally increased when not adjusted for inflation but decreased when adjusted for inflation, according to our analysis of USDA data.

The average per-acre payment for the overall program increased from $42.99 in fiscal year (FY) 1986 to $78.23 in FY 2023. Similarly, average per-acre rental payments for general enrollment increased from $42.99 in FY 1986 to $57.44 in FY 2023. In contrast, average per-acre rental payments for grassland enrollment decreased since its establishment, from $21.03 in FY 2016 to $15.57 in FY 2023. (See fig. 10.)

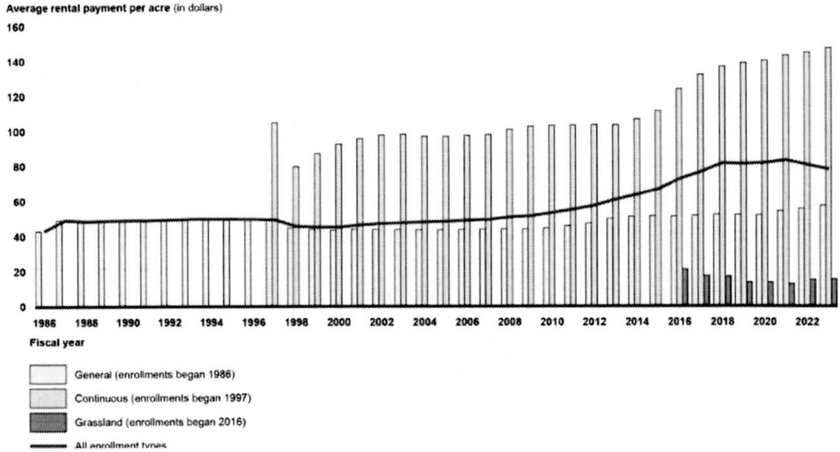

Source: GAO analysis of U.S. Department of Agriculture (USDA) information. | GAO-24-106311.

Note: To calculate the average rental rate, we averaged the contract rental rates, weighted by the number of acres under contract. This provided the rent paid for the average acre.

Figure 10. Average Rental Payment per Acre (not adjusted for inflation) for USDA's Conservation Reserve Program, by Type of Enrollment, Fiscal Years 1986–2023.

When adjusted for inflation, the average rental payment per acre decreased for overall and general enrollment. Specifically, the per-acre payment decreased from $100.44 in fiscal year 1986 to $78.23 for the program overall and $57.44 for general enrollment in FY 2023. For grassland enrollment, the decrease was greater when adjusting for inflation, from $25.59 in FY 2016 to $15.57 in FY 2023. (See fig. 11.)

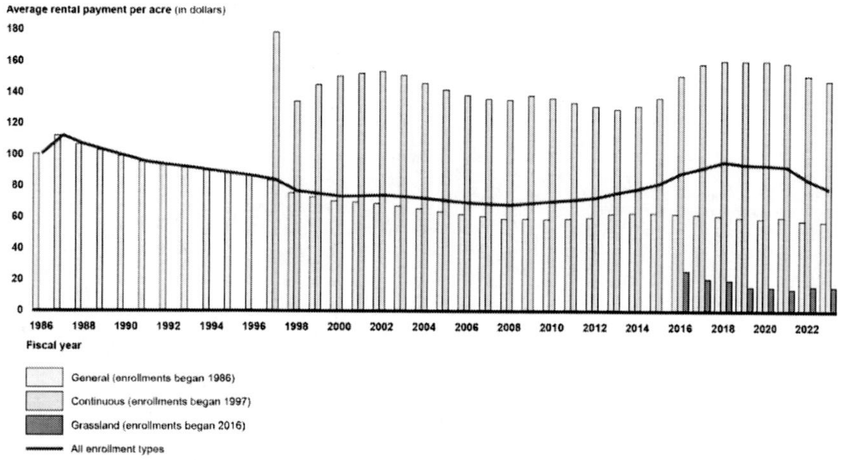

Source: GAO analysis of U.S. Department of Agriculture (USDA) information. | GAO-24-106311.

To calculate the average rental rate, we averaged the contract rental rates, weighted by the number of acres under contract. This provided the rent paid for the average acre.

We adjusted the rental payments for inflation to fiscal year 2023 dollars.

Figure 11. Average Rental Payment per Acre (adjusted for inflation) in USDA's Conservation Reserve Program, by Type of Enrollment, Fiscal Years 1986–2023.

Appendix II: Information on the Number of Years Land Is Enrolled

The length of contracts, years enrolled, and type of land enrolled in the Conservation Reserve Program (CRP), administered by the U.S. Department of Agriculture's (USDA) Farm Service Agency (FSA), have varied over time, according to our analysis of data from USDA.

Length of contracts. In general, the proportion of acres held in longer-term contracts has increased over time since fiscal year (FY) 1995, the earliest time frame for which data were available, according to our analysis of contracts for all three types of CRP enrollment (general, continuous, and grassland). For example, in FY 1995, over 99 percent of acres (34.8 million) were held in 10-year contracts, while 0.5 percent were held in 15-year contracts. In comparison, in FY 2023, just over half (51.9 percent or 11.9 million acres) were held in 10-year contracts, while 42.7 percent were held in in 15-year contracts. An additional 5.2 percent (1.9 million acres) were in 11- to 14-year contracts, and 0.2 percent were in 30-year contracts (see fig. 12).

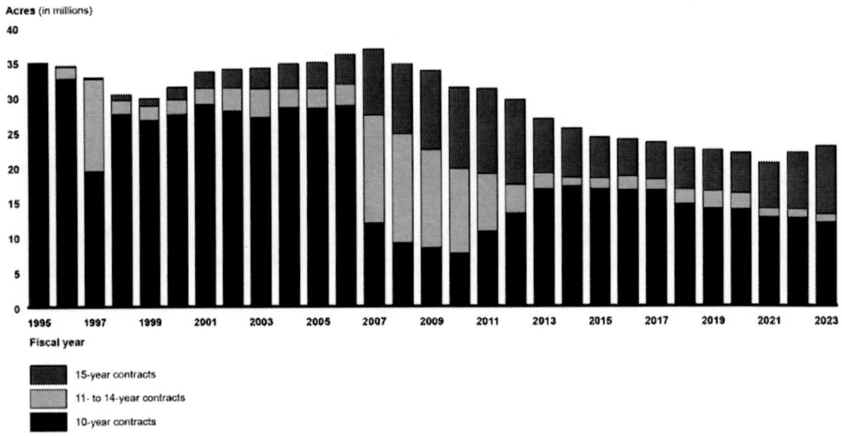

Source: GAO analysis of U.S. Department of Agriculture (USDA) information. | GAO-24-106311.

Note: General, continuous, and grassland enrollments began in 1986, 1997, and 2016, respectively. Contract lengths are generally 10 to 15 years, but contract lengths for certain initiatives can be as short as 3 years or as long as 30 years. There are so few contracts for 3 to 5 years and for 30 years that they are not visible in the figure.

Figure 12. Length of Contracts for Acres Enrolled in USDA's Conservation Reserve Program, Fiscal Years 1995–2023.

In 1994 and 1995, USDA offered 1-year contract extensions instead of conducting general sign-ups, according to a USDA FY 2019 annual summary for CRP.[64] These contract extensions help explain the increased portion of contracts in FY 1997 being longer than 10 years, according to an FSA official. Then, in part due to a high number of contracts due to expire, in 2007 through

[64] U.S. Department of Agriculture, Conservation Reserve Program: Annual Summary and Enrollment Statistics (FY 2019) (n.d.).

2010, USDA offered 2- to 5-year contract extensions, in part to stagger expirations over time, according to a 2017 study.[65] Landowners and operators holding 82 percent of expiring acreage opted for these extensions, according to that study.

Length of time enrolled. More than half of the 22.9 million acres enrolled in CRP as of September 30, 2023, had been in the program for less than 10 years, and more than 20 percent of the 22.9 million acres have been in CRP for at least 20 years, according to our analysis of USDA field-level data. Land can be re-enrolled after contracts expire, which explains why land may be enrolled for more than 10 or 15 years. Table 3 provides information on how long acres had been enrolled in the program as of the end of fiscal year 2023.

Table 3. Number of acres and years enrolled in USDA's Conservation Reserve Program (CRP), as of fiscal year 2023

Years enrolled in CRP	Acres (in millions)	Percentage of CRP total acres
1–10	13.9	61%
11–20	4.3	19%
21–30	2.3	10%
31 or more	2.5	11%
Total enrolled as of Sept. 30, 2023	22.9	100%

Source: GAO analysis of U.S. Department of Agriculture (USDA) information. | GAO-24-106311.

Note: The sum of the acres or percentages do not add to the totals due to rounding.

Appendix III: Information on How Acres and Rental Payments Are Distributed, by State

In fiscal year (FY) 2023, 22.9 million acres were enrolled in the Conservation Reserve Program (CRP), administered by the U.S. Department of Agriculture's (USDA) Farm Service Agency (FSA). Over half of these acres were in seven states in the central U.S.—Colorado, Iowa, Kansas, Nebraska, North Dakota, South Dakota, and Texas. Each of these seven states enrolled at least 1 million acres, and in total they had over 13 million CRP acres, or 57 percent of total CRP acreage. The states with the most acres varied by type of

[65] Daniel Hellerstein, "The US Conservation Reserve Program: The evolution of an enrollment mechanism," *Land Use Policy,* vol. 63 (2017): 601–610.

enrollment, as table 4 and figure 13 show. The states with the most acres in general enrollment were the Great Plains states of Colorado, Kansas, and Texas.[661] Continuous enrollment was most concentrated in the Dakotas, Illinois, Iowa, and Minnesota. Grassland enrollment was most concentrated in the Great Plains states of Colorado, Nebraska, New Mexico, and South Dakota.

Table 4. Five states with the highest total acreage enrolled in USDA's Conservation Reserve Program (CRP), by type of enrollment, fiscal year 2023

Acres in millions

General enrollment		Continuous enrollment		Grassland enrollment		Total CRP enrollment	
State	Acres	State	Acres	State	Acres	State	Acres
TX	1.5	IA	1.3	NE	1.2	CO	2.4
CO	1.1	SD	0.8	SD	1.2	TX	2.2
KS	1.1	MN	0.7	CO	1.2	SD	2.1
WA	0.7	ND	0.7	NM	0.7	NE	1.8
IA	0.3	IL	0.6	KS	0.3	KS	1.7

Source: GAO analysis of U.S. Department of Agriculture (USDA) information. | GAO-24-106311.

Total rental payments vary across states but do not necessarily correlate with acreage. Four states—Illinois, Iowa, Minnesota, and South Dakota—received at least $100 million each in rental payments, comprising almost half (48 percent) of total rental payments. The states with the highest rental payments varied by type of enrollment, though, as table 5 and figure 14 show. The five states receiving the most in total CRP rental payments are also the five states receiving the most in continuous enrollment rental payments—Illinois, Iowa, Minnesota, Missouri, and South Dakota—but those states varied with general and grassland enrollment.

66 The Great Plains in the U.S. include at least portions of 10 states: Colorado, Kansas, Montana, New Mexico, Nebraska, North Dakota, Oklahoma, South Dakota, Texas, and Wyoming.

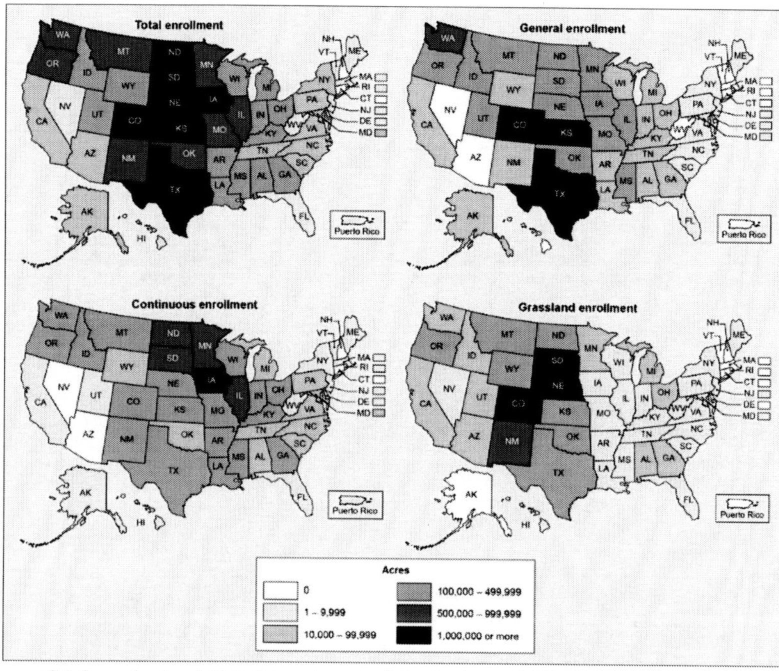

Source: GAO analysis of U.S. Department of Agriculture (USDA) information. | GAO-24-106311.

Figure 13. Total Acres Enrolled, by Type of Enrollment and State or Territory, in USDA's Conservation Reserve Program, Fiscal Year 2023.

Similar to total rental payments per state, rental payments per acre vary across states and do not necessarily correlate with acreage. Table 6 shows FY 2023 CRP rental payments per acre for the top five states, by each type of enrollment and total CRP enrollment. Figure 15 shows per- acre rental payments, by state and type of enrollment.

Table 5. Five states with the highest total rental payments in USDA's Conservation Reserve Program (CRP), by type of enrollment, fiscal year 2023

Dollars in millions

General enrollment		Continuous enrollment		Grassland enrollment		Total CRP enrollment	
State	Rental payments	State	Rental payments	State	Rental payments	State	Rental payments
IA	$60.3	IA	$342.0	SD	$19.2	IA	$402.5

Table 5. (Continued)

General enrollment		Continuous enrollment		Grassland enrollment		Total CRP enrollment	
TX	53.6	IL	140.5	CO	19.2	IL	173.7
KS	48.9	MN	128.8	NE	18.3	MN	150.8
MO	42.1	SD	103.8	NM	10.2	SD	129.5
WA	39.8	MO	57.7	KS	5.6	MO	99.8

Source: GAO analysis of U.S. Department of Agriculture (USDA) information. | GAO-24-106311.

Note: Rental payments do not include certain cost-share payments (up to 50 percent of the reimbursable cost of installing a conservation practice) and incentive payments (incentives and adjustments such as a sign-up incentive payment of 32.5 percent of the first full year annual rental payment at contract approval and incentives tied to specific conservation practices).

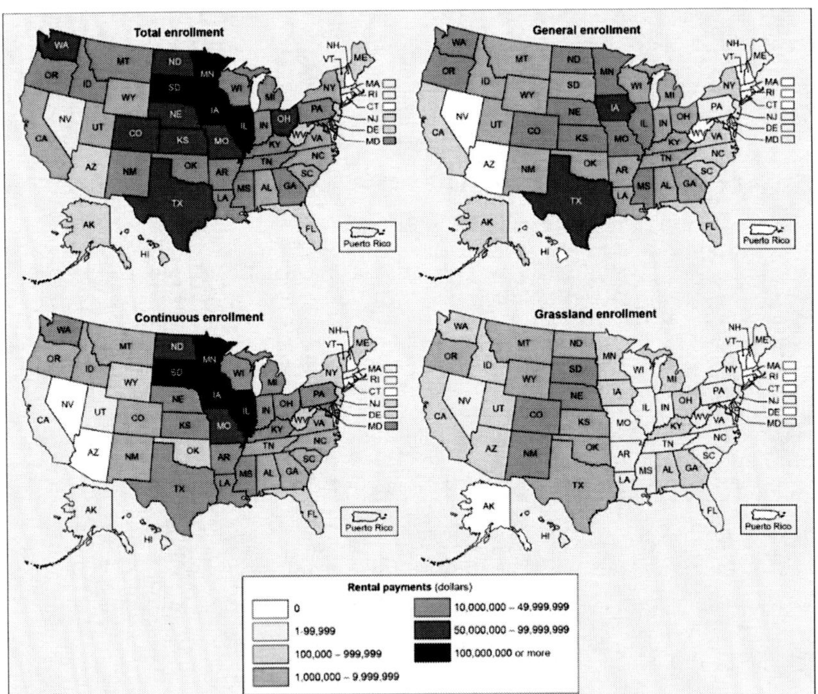

Source: GAO analysis of U.S. Department of Agriculture (USDA) information; Map Respources (maps) | GAO-24-106311.

Figure 14. Total Rental Payments, by Type of Enrollment and State or Territory, for USDA's Conservation Reserve Program, Fiscal Year 2023.

Table 6. Five states with the highest average rental payments per acre in USDA's Conservation Reserve Program (CRP), by type of enrollment, fiscal year 2023

In dollars

General enrollment		Continuous enrollment		Grassland enrollment		Total CRP enrollment	
State	Average rental rate	State	Average rental rate	State	Average rental rate	State	Average rental rate
IA	$173	IA	$256	IA	$38	MD	$243
IL	156	OH	252	MD	34	IA	238
IN	136	MD	250	IN	34	OH	223
KY	135	IL	234	IL	30	IL	214
MO	132	IN	223	WI	27	IN	207

Source: GAO analysis of U.S. Department of Agriculture (USDA) information. | GAO-24-106311.

Note: Rental payments do not include certain cost-share payments (up to 50 percent of the reimbursable cost of installing a conservation practice) and incentive payments (incentives and adjustments such as a sign-up incentive payment of 32.5 percent of the first full year annual rental payment at contract approval and incentives tied to specific conservation practices).

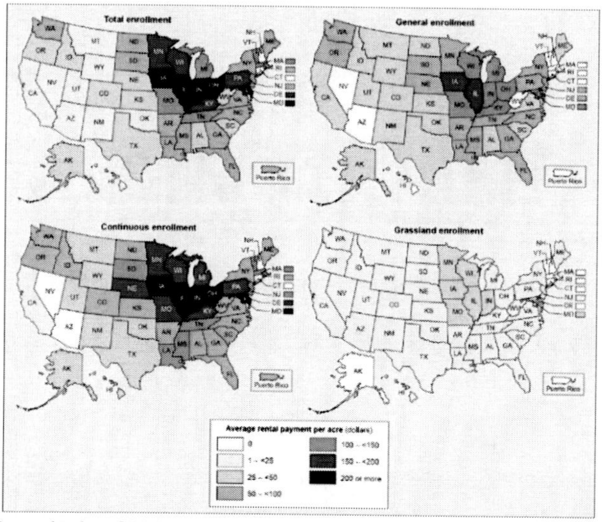

Source: GAO analysis of U.S. Department of Agriculture (USDA) information; Map Respources (maps) | GAO-24-106311.

Figure 15. Average Rental Payment per Acre for USDA's Conservation Reserve Program, by Type of Enrollment and State or Territory, Fiscal Year 2023.

Table 7. Total acres enrolled and total rental payments in USDA's Conservation Reserve Program (CRP), by type of enrollment and by state or territory, fiscal year 2023

General enrollment			Continuous enrollment		Grassland enrollment		Total CRP enrollment	
State or territory	Acres	Rental payment	Acres	Rental payment	Acres	Rental payment	Acres	Rental payment
AK	10.8k	$444.1k	693	$31.7k	0	$0	11.5k	$475.9k
AL	61.4k	$3.0m	33.7k	$1.7m	11.9k	$192.5k	$107.0k	$4.9m
AR	25.6k	$1.8m	166.9k	$14.5m	452	$6.5k	193.0k	$16.4m
AZ	0	$0	0	$0	15.2k	$208.2k	15.2k	$208.2k
CA	23.2k	$835.1k	1.0k	$25.9k	39.6k	$595.0k	63.8k	$1.5m
CO	1.1m	$34.0m	128.1k	$8.2m	1.2m	$19.2m	2.4m	$61.3m
CT	0	$0	0	$0	0	$0	0	$0
DE	80	$6.7k	3.0k	$466.6k	0	$0	3.0k	$473.3k
FL	4.1k	$274.0k	2.1k	$132.5k	1.5k	$29.4k	7.7k	$435.9k
GA	31.9k	$2.4m	110.3k	$8.4m	17.6k	$371.4k	159.8k	$11.1m
HI	0	$0	1.3k	$61.6k	3.7k	$40.7k	5.0k	$102.3k
IA	348.9k	$60.3m	1.3m	$342.0m	3.8k	$145.2k	1.7m	$402.5m
ID	183.9k	$8.4m	154.9k	$9.4m	38.2k	$426.3k	377.0k	$18.3m
IL	212.0k	$33.2m	600.1k	$140.5m	1.0k	$31.4k	813.1k	$173.7m
IN	34.4k	$4.7m	157.3k	$35.0m	294	$9.9k	192.0k	$39.7m
KS	1.1m	$48.9m	349.4k	$21.3m	331.4k	$5.6m	1.8m	$75.8m
KY	42.7k	$5.8m	100.8k	$18.9m	5.2k	$127.8k	148.7k	$24.8m
LA	27.2k	$1.8m	223.1k	$23.3m	1.4k	$26.5k	251.7k	$25.1m
MA	0	$0	9	$865	0	$0	9	$865
MD	625	$67.3k	41.3k	$10.3m	1.1k	$36.3k	43.0k	$10.4m
ME	1.2k	$51.8k	1.9k	$137.3k	89	$1.4k	3.2k	$190.4k
MI	20.4k	$2.2m	67.4k	$11.4m	26.4k	$435.3k	114.2k	$14.1m
MN	222.8k	$21.6m	741.8k	$128.8m	11.5k	$300.5k	976.1k	$150.8m

General enrollment			Continuous enrollment		Grassland enrollment		Total CRP enrollment	
State or territory	Acres	Rental payment	Acres	Rental payment	Acres	Rental payment	Acres	Rental payment
MO	319.8k	$42.1m	381.0k	$57.7m	3.4k	$87.0k	704.2k	$99.8m
MS	184.6k	$12.2m	285.3k	$29.3m	849	$12.3k	470.8k	$41.5m
MT	310.5k	$9.2m	178.0k	$6.9m	319.2k	$4.0m	807.7k	$20.0m
NC	9.0k	$574.0k	11.7k	$1.2m	853	$16.2k	21.5k	$1.7m
ND	286.4k	$13.5m	733.6k	$53.9m	104.6k	$1.7m	1.1m	$69.1m
NE	347.0k	$23.1m	268.2k	$43.4m	1.2m	$18.3m	1.8m	$84.9m
NH	0	$0	0	$0	0	$0	0	$0
NJ	144	$7.8k	1.3k	$141.3k	173	$3.5k	1.7k	$152.6k
NM	88.8k	$2.4m	109.0k	$3.2m	659.2k	$10.2m	857.0k	$15.8m
NV	0	$0	0	$0	3.0k	$30.0k	3.0k	$30.0k
NY	2.6k	$149.6k	9.5k	$947.2k	3.4k	$72.2k	15.5k	$1.2m
OH	27.3k	$3.2m	187.0k	$47.2m	12.7k	$302.2k	227.0k	$50.7m
OK	307.2k	$8.3m	12.7k	$517.8k	303.2k	$5.3m	623.1k	$14.1m
OR	348.1k	$19.0m	105.3k	$8.2m	146.3k	$1.8m	599.7k	$29.0m
PA	133	$8.8k	81.4k	$12.7m	964	$22.9k	82.5k	$12.7m
PR	0	$0	495	$25.4k	0	$0	495	$25.4k
RI	0	$0	28	$2.1k	0	$0	28	$2.1k
SC	4.8k	$184.3k	20.8k	$1.1m	1.8k	$35.7k	27.4k	$1.3m
SD	100.7k	$6.6m	808.8k	$103.8m	1.2m	$19.2m	2.1m	$129.5m
TN	47.5k	$4.6m	47.6k	$5.8m	3.8k	$54.4k	98.8k	$10.4m
TX	1.5m	$53.6m	315.7k	$13.5m	317.2k	$5.4m	2.2m	$72.5m
UT	101.7k	$2.9m	213	$9.3k	56.0k	$564.8k	157.8k	$3.5m
VA	6.1k	$254.3k	18.0k	$1.5m	5.4k	$94.5k	29.5k	$1.9m
VT	0	$0	2.0k	$240.2k	0	$0	2.0k	$240.2k
WA	674.4k	$39.8m	275.0k	$21.6m	19.7k	$226.8k	969.2k	$61.6m
WI	57.9k	$6.9m	134.4k	$26.6m	1.1k	$30.2k	193.4k	$33.6m
WV	0	$0	4.7k	$394.2k	5.0k	$63.0k	9.6k	$457.3k

Table 7. (Continued)

WY	82.4k	$2.1m	10.4k	$299.5k	283.3k	$3.6m	376.1k	$6.0m
Total	8.4m	$480.5m	8.2m	$1,214.8m	6.3m	$98.8m	22.9m	$1,794.1m

Legend: k = thousands; m = millions.

Source: GAO analysis of U.S. Department of Agriculture (USDA) information. | GAO-24-106311.

Note: The sum of the acres or rental payments may not add to the totals due to rounding.

Rental payments do not necessarily correlate to acreage because the program's rental rates are higher in some parts of the country than others and vary by type of enrollment. FSA calculates a CRP rental payment per acre for general and continuous enrollment using per-acre estimates of county average market dryland and irrigated cash rental rates for cropland from USDA's National Agricultural Statistics Service (NASS). FSA also uses additional information supplied by state FSA state committees and Conservation Reserve Enhancement Program partners. These rental rates are generally higher in Illinois, Indiana, and Iowa, compared with states farther West, which have more enrolled acres. For grassland enrollment, the rental payment per acre is based on the pastureland rental rates from NASS surveys and are considerably less than for cropland.

Table 7 provides information by state on the number of acres and rental payments for each type of enrollment and in total for CRP.

Appendix IV: Information on Acres Enrolled, by County, Fiscal Years 2003, 2013, and 2023

This appendix provides county-level information on acres enrolled in the Conservation Reserve Program (CRP), administered by the U.S. Department of Agriculture's (USDA) Farm Service Agency, for select fiscal years (FY). Overall, the number of acres enrolled in CRP have fluctuated over time. In addition, the portion of land in continuous and grassland enrollments have increased since each type was introduced, while acres in general enrollment have decreased. Figures 16 through 18 show county-level information on acres in the program for FYs 2003, 2013, and 2023, by type of enrollment, according to our analysis of USDA data.

In FY 2003, CRP enrolled a total of 34.1 million acres. The acres in general enrollment (31.6 million) were most concentrated in western Kansas, the Texas panhandle, eastern Colorado, and North Dakota with other areas of substantial acres, such as southern Iowa, northern Missouri, and eastern Washington. Continuous enrollment acres (2.5 million) were most concentrated in Iowa, Illinois, and western and southern Minnesota. (See fig. 16.)

In FY 2013, CRP enrolled a total of 26.8 million acres, a decrease of about 7.3 million acres from FY 2003. Acres in general enrollment (21.4 million) also decreased from FY 2003, while continuous enrollment acres (5.5 million) more than doubled. The decrease in general enrollment acres was broadly dispersed and included counties in Montana, North Dakota, and South Dakota. The increase in continuous enrollment acres included counties in North Dakota, South Dakota, Kansas, Iowa, Illinois, eastern Arkansas, eastern Louisiana, Mississippi, southern Georgia, and Pennsylvania. (See fig. 17.)

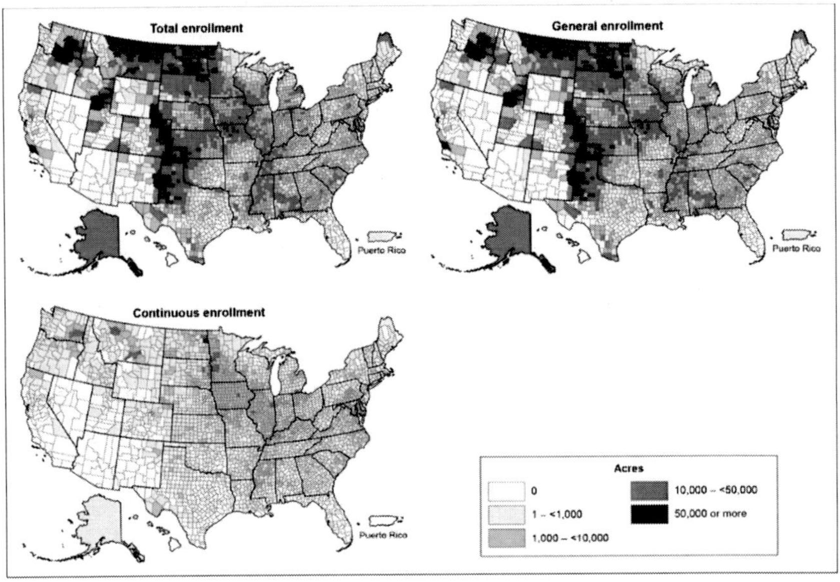

Source: GAO analysis of U.S. Department of Agriculture (USDA) data, Census Bureaus (georgraphic boudaries) | GAO-24-106311.

Note: General, continuous, and grassland enrollments began in 1986, 1997, and 2016, respectively. As a result, grassland enrollment is not included in this figure. All values represent county-level information for the Conservation Reserve Program, except for Alaska and Puerto Rico, which are at the state (territory) level.

Figure 16. Acres Enrolled in USDA's Conservation Reserve Program, by Type of Enrollment and County, Fiscal Year 2003.

In FY 2023, CRP enrolled a total of 22.9 million acres, a decrease of about 4 million acres from FY 2013. Acres in general enrollment (8.4 million) also decreased from FY 2013, while acres in continuous enrollment (8.2 million) increased. Grassland enrollment acres, which were zero in 2013, represented over a quarter (6.3 million) of total acres in FY 2023. The decreases in general enrollment acres occurred broadly, including in in Montana, North Dakota, South Dakota, and western Kansas, as well as other areas, such as southern

Iowa, northern Missouri, and Mississippi. The increase in continuous enrollment acres included counties in Iowa, North Dakota, South Dakota, western Kansas, the Texas panhandle, and eastern New Mexico. Grassland enrollment acres were mostly in counties in a band stretching from western South Dakota to eastern New Mexico, along with counties in other states, including Montana, Wyoming, and Oregon. (See fig. 18.)

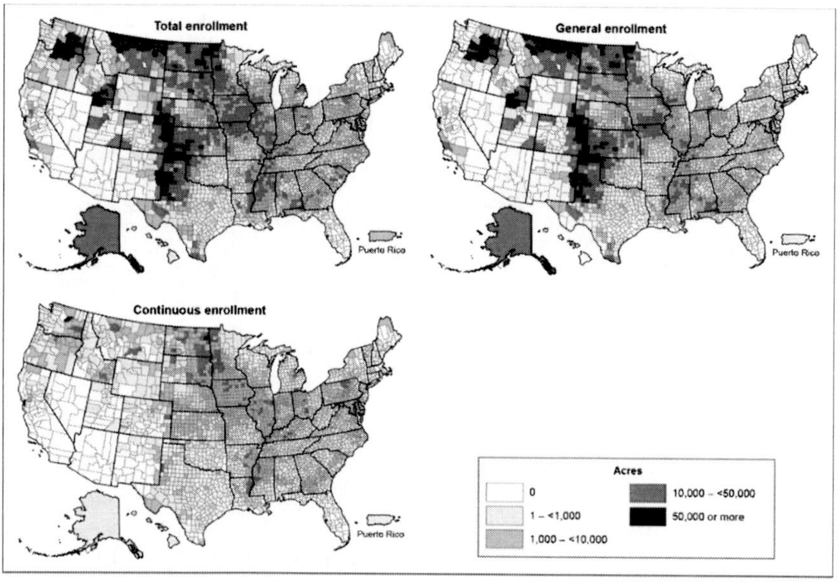

Source: GAO analysis of U.S. Department of Agriculture (USDA) data, Census Bureaus (georgraphic boudaries) | GAO-24-106311.

Note: General, continuous, and grassland enrollments began in 1986, 1997, and 2016, respectively. As a result, grassland enrollment is not included in this figure. All values represent county-level information for the Conservation Reserve Program, except for Alaska and Puerto Rico, which are at the state (territory) level.

Figure 17. Acres Enrolled in USDA's Conservation Reserve Program, by Type of Enrollment and County, Fiscal Year 2013.

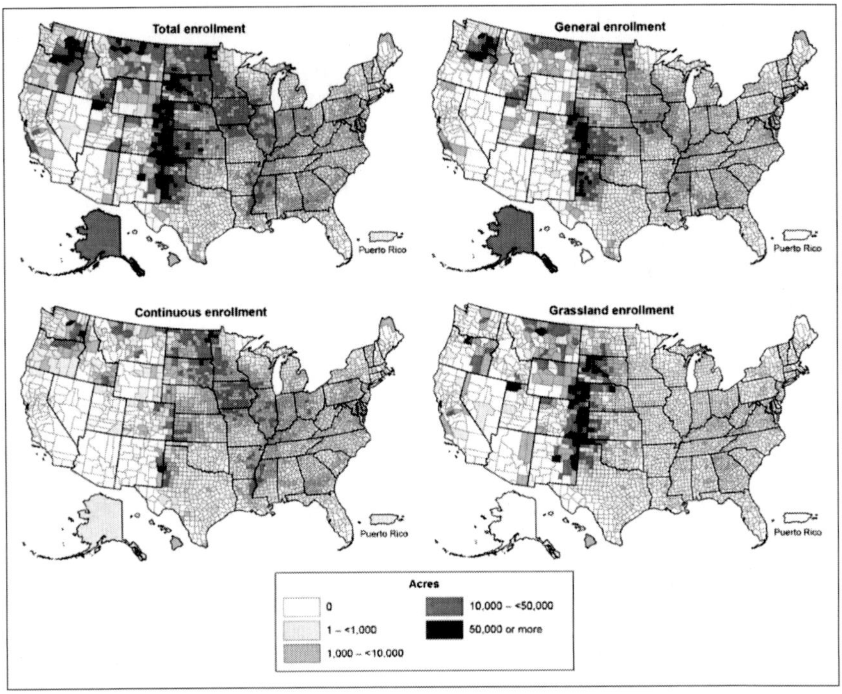

Source: GAO analysis of U.S. Department of Agriculture (USDA) data, Census Bureaus (georgraphic boudaries) | GAO-24-106311.

Note: All values represent county-level information for the Conservation Reserve Program, except for Alaska and Puerto Rico, which are at the state (territory) level.

Figure 18. Acres Enrolled in USDA's Conservation Reserve Program, by Type of Enrollment and County, Fiscal Year 2023.

Appendix V: Information on Rental Payments for Acres Enrolled, by County, Fiscal Year 2023

We analyzed rental payments by county in total and by enrollment type for the Conservation Reserve Program (CRP), administered by the U.S. Department of Agriculture's (USDA) Farm Service Agency, for fiscal year (FY) 2023, using USDA data (see fig. 19). In FY 2023, USDA made a total of $1.8 billion in annual rental payments. Specifically, information for each enrollment type is as follows:

- For general enrollment, rental payments were $481 million (27 percent of the total). Counties with relatively high rental payments were in Washington and Oregon as well as a band of counties from eastern Colorado and western Kansas to the Texas panhandle.
- For continuous enrollment, rental payments were $1.2 billion (68 percent of the total). Counties with relatively high rental payments extend from eastern North Dakota and western Minnesota through eastern South Dakota and Iowa to Illinois.
- For grassland enrollment, rental payments were $99 million (6 percent of the total). Counties with relatively high rental payments were in the Great Plains in western South Dakota, western Nebraska, eastern Colorado, and eastern New Mexico.

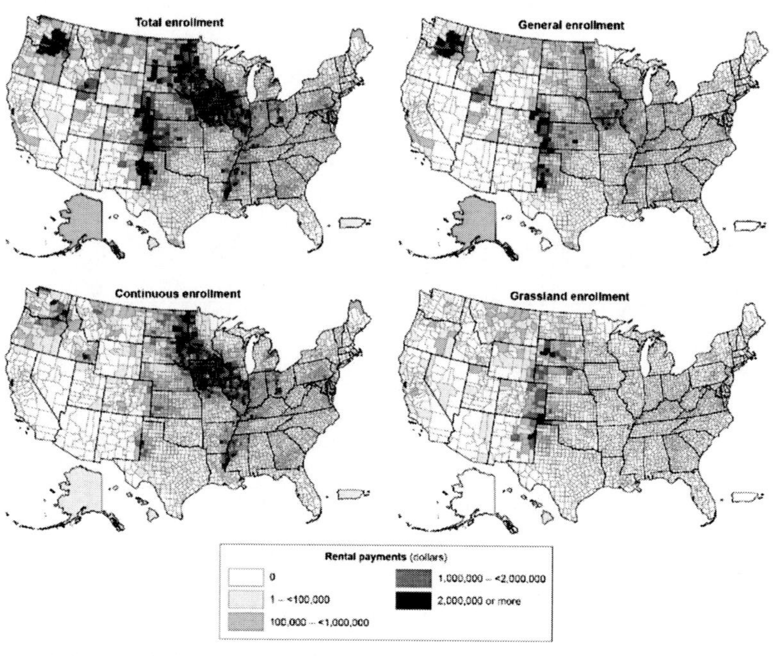

Source: GAO analysis of U.S. Department of Agriculture (USDA) data, Census Bureaus (georgraphic boudaries) | GAO-24-106311.

Note: All values represent county-level information for the Conservation Reserve Program, except for Alaska and Puerto Rico, which are at the state (territory) level.

Figure 19. Rental Payments for Acres Enrolled in USDA's Conservation Reserve Program, by Type of Enrollment and County, Fiscal Year 2023.

Appendix VI: Information on Conservation Practices, by Acreage and Rental Payments, Fiscal Year 2023

To participate in the Conservation Reserve Program (CRP), landowners and operators devote land to various conservation practices to help achieve long-term program goals such as preventing soil erosion, improving water quality, and reducing loss of wildlife habitat. Three practices related to the establishment of grasses or legumes accounted for over 60 percent of the land enrolled in CRP in fiscal year (FY) 2023, as shown in table 8. The next largest conservation practices in terms of acreage were the establishment or maintenance of permanent wildlife habitat and restoration of rare and declining habitat.[78] Table 8 shows the largest conservation practices in terms of acreage, by type of enrollment, in CRP, administered by the U.S. Department of Agriculture's Farm Service Agency, in FY 2023.

Table 8. Largest conservation practices, by acreage and type of enrollment, for USDA's Conservation Reserve Program (CRP), fiscal year 2023

Acres (in millions) (percentage)				
Conservation practice	General enrollment	Continuous enrollment	Grassland enrollment	Total CRP enrollment
Maintenance of permanent grass and legumes on grassland[a]	0.0 (0%)	0.0 (0%)	6.2 (98%)	6.2 (27%)
Establishment or maintenance of permanent native grasses	3.7 (45%)	1.2 (14%)	0.0 (0%)	4.9 (21%)
Establishment or maintenance of permanent introduced grasses and legumes	2.1 (26%)	1.1 (14%)	0.0 (0%)	3.3 (14%)
Establishment or maintenance of permanent wildlife habitat	0.9 (11%)	0.8 (10%)	0.0 (0%)	1.7 (7%)
Restoration of rare and declining habitat[b]	1.0 (12%)	0.3 (3%)	0.0 (0%)	1.3 (6%)
Restoration of wetlands (non-floodplain)	0.0 (0%)	0.8 (10%)	0.0 (0%)	0.8 (4%)
Establishment or maintenance of filter strips[c]	0.0 (0%)	0.6 (8%)	0.0 (0%)	0.6 (3%)
Establishment or maintenance of riparian buffers[d]	0.0 (0%)	0.5 (6%)	0.0 (0%)	0.5 (2%)
Establishment or maintenance of pollinator habitat	0.1 (1%)	0.5 (6%)	0.0[e] (0%)	0.5 (2%)
Restoration of wetlands on floodplains	0.0[f] (1%)	0.4 (5%)	0.0 (0%)	0.5 (2%)

[78] This practice is intended to restore critically endangered, endangered, and threatened habitats.

Table 8. (Continued)

Acres (in millions) (percentage)				
Conservation practice	General enrollment	Continuous enrollment	Grassland enrollment	Total CRP enrollment
Other	0.5 (6%)	2.1 (25%)	0.1 (2%)	2.6 (11%)
Total	8.4 (100%)	8.2 (100%)	6.3 (100%)	22.9 (100%)

Source: GAO analysis of U.S. Department of Agriculture (USDA) information. | GAO-24-106311

Notes: Percentage refers to the percentage of acres specific to each enrollment type—general, continuous, or grassland. For total CRP enrollment, percentage refers to acres for the entire program.

Legumes are plants that have seeds in a pod and improve soil by taking nitrogen from the air.

Wetlands are low-lying ecosystems where the water table is always at or near the surface. They perform vital ecological functions, including providing critical habitat for wildlife and waterfowl, mitigating floods by slowing down and absorbing excess water during storms, and purifying water by filtering out pollutants before they enter streams and lakes, among other things.

The sum of the acres or percentages may not add to the totals due to rounding.

[a] This practice applies only to grassland enrollment and the land may be used for haying or grazing.

[b] This practice is intended to restore critically endangered, endangered, and threatened habitats.

[c] A filter strip is a narrow band of grasses, legumes, and forbs—plants that are not woody and have broad leaves. Filter strips improve water quality by intercepting sediment and nutrients, mitigate erosion by reducing the negative impacts of wind and water, and provide habitat and corridors for wildlife.

[d] Riparian buffers are strips of trees bordering streams, wetland areas, and other waterbodies to improve water quality and wildlife habitat.

[e] Grassland enrollment, in fiscal year 2023, included 3,100 acres of pollinator habitat, which rounds to 0.0 in the table.

[f] General enrollment, in fiscal year 2023, included 48,630 acres of wetland restoration on floodplains, which rounds to 0.0 in the table.

The three practices related to establishment of grasses or legumes accounted for about 41 percent of rental payments in FY 2023, as shown in table 9. Establishment or maintenance of filter strips and restoration of rare and declining habitat were the next largest conservation practices in terms of rental payments.[79] Table 9 shows the largest conservation practices in terms of rental payments, by type of enrollment, in CRP, in FY 2023.

[79] A filter strip is a narrow band of grasses, legumes, and forbs—plants that are not woody and have broad leaves. Filter strips improve water quality by intercepting sediment and nutrients, mitigate erosion by reducing the negative impacts of wind and water, and provide habitat and corridors for wildlife.

Table 9. Largest conservation practices, by total rental payments, for USDA's Conservation Reserve Program (CRP), fiscal year 2023

Rental payments (dollars in millions) (percentage)				
Conservation practice	General enrollment	Continuous enrollment	Grassland enrollment	Total CRP enrollment
Establishment or maintenance of permanent native grasses	$172.1 (36%)	$140.6 (12%)	$0.0 (0%)	$312.8 (17%)
Establishment or maintenance of permanent introduced grasses and legumes	$133.5 (28%)	$142.0 (12%)	$0.0 (0%)	$275.5 (15%)
Establishment or maintenance of permanent wildlife habitat	$51.8 (11%)	$100.8 (8%)	$0.0 (0%)	$152.6 (9%)
Establishment or maintenance of filter strips[a]	$0.0 (0%)	$142.7 (12%)	$0.0 (0%)	$142.7 (8%)
Restoration of rare and declining habitat[b]	$77.1 (16%)	$56.5 (5%)	$0.0 (0%)	$133.6 (7%)
Restoration of wetlands (non-floodplain)	$0.0 (0%)	$117.4 (10%)	$0.0 (0%)	$117.4 (7%)
Establishment or maintenance of pollinator habitat	$6.8 (1%)	$105.3 (9%)	$0.1 (0%)	$112.2 (6%)
Maintenance of permanent grass and legumes on grassland	$0.0 (0%)	$0.0 (0%)	$96.7 (98%)	$96.7 (5%)
Wetland restoration on floodplains	2.9 (1%)	$76.6 (6%)	$0.0 (0%)	$79.5 (4%)
Establishment or maintenance of riparian buffers[c]	$0.0 (0%)	$68.2 (6%)	$0.0 (0%)	$68.2 (4%)
Other	$36.3 (8%)	$264.6 (22%)	$2.1 (2%)	$303.0 (17%)
Total	$480.5 (100%)	$1,214.8 (100%)	$98.8 (100%)	$1,794.1 (100%)

Source: GAO analysis of U.S. Department of Agriculture (USDA) information. | GAO-24-106311.

Notes: Percentage refers to the percentage of rental payments specific to each enrollment type—general, continuous, or grassland. For total CRP enrollment, percentage refers to rental payments for the entire program.

Legumes are plants that have seeds in a pod and improve soil by taking nitrogen from the air.

Wetlands are low-lying ecosystems where the water table is always at or near the surface. They perform vital ecological functions, including providing critical habitat for wildlife and waterfowl, mitigating floods by slowing down and absorbing excess water during storms, and purifying water by filtering out pollutants before they enter streams and lakes, among other things.

The sum of the rental payments or percentages may not add to the totals due to rounding.

[a] A filter strip is a narrow band of grasses, legumes, and forbs—plants that are not woody and have broad leaves. Filter strips improve water quality by intercepting sediment and nutrients, mitigate erosion by reducing the negative impacts of wind and water, and provide habitat and corridors for wildlife.

[b] This practice is intended to restore critically endangered, endangered, and threatened habitats.

[c] Riparian buffers are strips of trees bordering streams, wetland areas, and other waterbodies to improve water quality and wildlife habitat.

Appendix VII: Methodology for Literature Review and Expert Selection

We conducted our review of literature related to the Conservation Reserve Program (CRP) from January 2023 to August 2023. To identify studies on CRP and the methods that the U.S. Department of Agriculture's (USDA) Farm Service Agency (FSA) uses to select CRP offers, we searched publications from the USDA's Economic Research Service (ERS) website since 1999, as well as journals published by the Agricultural and Applied Economics Association. We also searched databases, including Agricola, Econlit, and Scopus for publications since 1996, using the keywords "conservation reserve program" and "cost- benefit," "environmental benefit index," and "cost-effectiveness." Additionally, we identified related studies from these searches and also from the experts we interviewed. Our search resulted in identifying 57 studies, including peer-reviewed journal articles and USDA reports.

We evaluated these 57 studies and selected 34 of them based on their methodological rigor and relevancy. All 34 studies were published in peer-reviewed journals or by USDA. To assess the methodological quality of the selected studies, we obtained information about each study and the methodology, and we found that the data sources, modeling, and statistical analyses were clearly presented based on generally accepted social science standards. Relevancy was primarily based on the study examining the benefits and cost-effectiveness of CRP and identifying improvements that may be made in FSA's methods to score and select CRP offers.

The structure of the Environmental Benefits Index (EBI)—that is, adding environmental benefits factors to a cost factor—has not changed since 1997. Thus, we did not include studies prior to that date but included several studies after that date that are more than 10 years old. While the magnitude of estimated effects drawn from older studies may or may not apply to current conditions, their suggestions on how the EBI can be improved are still valid.

We also interviewed eight experts who primarily authored reports and articles identified through our literature searches. The experts were three academics, as well as three current and two former officials from USDA's ERS or FSA. These experts have knowledge of and experience with CRP, and with scoring and selecting CRP offers. They have generally published multiple articles on CRP's methods for scoring and selecting offers.

We asked these experts a variety of questions on FSA's methods, including some questions beyond the scope of their studies. Not every expert answered every question or commented on every topic, in part because they have different areas of expertise and experience.

Appendix VIII: Methodology for Data Analysis

To provide detailed information on the Conservation Reserve Program (CRP), administered by the U.S. Department of Agriculture's (USDA) Farm Service Agency, we analyzed USDA data from fiscal years (FY) 1986 through 2023, and presented the results of these analyses in objective one and appendixes I to VI. Specifically, we analyzed the following:

- acres enrolled in CRP and total rental payments, FY 1986 through FY 2023;
- average rental payments per acre, FY 1986 through FY 2023;[80]
- number of years land is enrolled in CRP, as of FY 2023;
- acres enrolled in CRP and total rental payments, by state, FY 2023;
- acres enrolled in CRP, by county, FY 2003, FY 2013, and FY 2023;
- total rental payments, by county, FY 2023; and
- largest conservation practices in terms of acreage and rental payments, FY 2023.

Our analysis described above was based on USDA's annual snapshots of its data on CRP from FY 1995 (the earliest available) through FY 2023, as well as number of acres enrolled and related USDA rental payments per fiscal year nationwide, which were publicly available on USDA's website. USDA's annual snapshots include contract-level data, such as data on CRP sign-ups, conservation practices, and geographic locations associated with the contracts.

USDA's rental payments for CRP land occur in arrears (e.g., contracted acres and rental rate in 2002 result in a rental payment in 2003). USDA has reported on rental payments using this method. However, rental payments in this report represent the amount obligated that fiscal year, based on CRP acres and rental rates. To effectively compare USDA's rental payments for CRP over time, we adjusted the payments using the Personal Consumption

[80] To calculate the average rental rate, or rental payment per acre, we averaged the contract rental rates, weighted by the number of acres under contract. This provided the rent paid for the average acre.

Expenditures price index. FSA officials agreed this was an appropriate price index to use for our analysis.

To assess the reliability of these data, we performed electronic data testing to identify missing values, unexpected values, omissions, logical inconsistencies, and duplicates. Before implementing data cleaning procedures, we checked the data's consistency against aggregated year- by-year CRP totals that FSA published at the county and state levels. We made small adjustments to the CRP data to correct observed anomalies. For example, we corrected invalid FIPS codes associated with the locations of a small number of contracts, and we removed a small number of duplicate contracts. The effect of these adjustments on our reported totals is minimal. Finally, we interviewed knowledgeable FSA staff about how to interpret certain data fields and about known data limitations. We concluded that the data were sufficiently reliable for the purposes of reporting CRP acreage and payment information.

Appendix IX: Factors and Related Points for the Environmental Benefits Index

The U.S. Department of Agriculture's Farm Service Agency uses an Environmental Benefits Index (EBI) to select land offered for general enrollment in its Conservation Reserve Program (CRP). The EBI is composed of five factors related to environmental benefits and a factor related to cost. Table 10 describes the factors, subfactors, and related points in the EBI.

Table 10. Factors and Related Points in the Environmental Benefits Index that USDA Uses in Its Conservation Reserve Program (CRP)

Factors and subfactors	Purpose	Possible points
Wildlife benefits	To encourage landowners and operators to plant cover on contract acreage that will be beneficial to wildlife. It consists of three subfactors.[a]	0 to 100
Cover	To evaluate cover based on the covers expected to become established and the planting rates.[b]	0 to 50
Wildlife enhancement	To evaluate enhancements based on the size, shape, and composition of the potential wildlife habitat or food plot.	0 to 20
Wildlife priority zone[c]	To identify locations of land that may contribute to the restoration of habitat of threatened or endangered species or contribute to the	0 or 30

Factors and subfactors	Purpose	Possible points
	restoration of certain species of national, regional, state, or local significance.	
Water quality benefits	To evaluate the potential for improving water quality by reducing (1) sediment, nutrients, and other pollutants from entering water ways and aquifers; (2) downstream flood damage through the restoration of hydrology; and (3) the leaching of nitrates and pesticides into groundwater.	0 to 100
Water quality zone[c]	To identify locations where proposed land may contribute to groundwater or surface water quality impairment and assist in meeting certain federal, state, or local water quality laws.	0 or 30
Groundwater quality	To evaluate the potential for the movement of nutrients and pesticides for a specific soil.	0 to 25
Surface water quality	To evaluate the potential amount of sediment that is delivered to water ways, the population within the watershed that would benefit most directly from improved surface water quality, and other related factors.	0 to 45
Soil erosion benefits	To evaluate the on-farm benefits of reduced soil erosion, including sheet and rill erosion, and wind erosion[d]	0 to 100
Enduring benefits	To evaluate the likelihood that the practice established will persist, be maintained, and provide benefits beyond the initial CRP contract period.	0 to 50
Air quality benefits	To evaluate the air quality improvements from reducing airborne dust and particulate from cropland wind erosion that causes damage to nearby affected population concentrations. It also provides points for the value of land for carbon sequestration.	3 to 45
Wind erosion impacts	To evaluate the potential wind erosion impacts, including the distance-weighted population that may be impacted by wind erosion.	0 to 25
Air quality zone[c]	To evaluate whether land is located in areas where agriculture significantly impacts air quality or is located within 50 miles of national parks with high-quality air standards.	0 or 10
Carbon sequestration	To provide a relative index of the projected carbon sequestration benefits of CRP cover types over the expected life of the practice.	3 to 10
Cost factor	To consider the relative cost of CRP offers by providing greater weight to offers with lower rental rates.	0 to 150
Cost subfactor	To determine the number of points to assign an offer based on the proposed rental rate (bid) and other factors. The formula is: a*(1-(Bid/b)), where "a" equals the maximum number of points the Farm Service Agency decides to assign to CRP offers for cost, and "b" is the maximum rental rate USDA will provide for CRP offers.	0 to 125
Offer less than the maximum rental rate	To provide points for offers whose offered rental rates are below the maximum per acre rental rate.	0 to 25

Source: GAO analysis of U.S. Department of Agriculture (USDA) information. | GAO-24-106311.

[a] Operators include those who lease the land or otherwise have the landowner's approval to enroll in the program.

[b] A planting rate is an expression of the number of pounds of seed per acre.

[c] Zones are defined by state committees, which are primarily active farmers and ranchers nominated by members of Congress and selected by the Secretary of Agriculture. The state committee may

consult with a state technical committee, which includes agricultural producers, nonindustrial private forest landowners, and other professionals who represent a variety of disciplines in soil, water, wetlands, plant, and wildlife sciences, including USDA and state representatives.

[d] Sheet and rill erosion is the physical removal of soil from the land surface by the action of rainfall, melting snow, irrigation, or runoff. Sheet erosion is a process in which detached soil is moved across the soil surface by sheet flow, often in early stages of runoff. Rill erosion occurs as runoff water begins to concentrate in small channels or streamlets. Wind erosion is the detachment and transport of soil particles caused by wind.

Appendix X: Factors and Related Points for the Grassland Index

The U.S. Department of Agriculture's Farm Service Agency uses an index to select land offered for grassland enrollment in its Conservation Reserve Program (CRP). The grassland index has seven ranking factors that total to determine the score. Four of these factors relate to environmental benefits, one relates to the demographic of the landowner or operator, one relates to the size of the livestock operation, and the seventh relates to the cost per acre of the offer. Table 11 describes the factors, subfactors, and related points for the index.

Table 11. Factors, subfactors, and related points of the grassland index USDA Uses for its Conservation Reserve Program (CRP)

Factors and subfactors	Purpose	Possible points
Current and future use	To encourage continued usage of the land and to retain most of the environmental benefits of expiring CRP land.	0 to 30
Expiring CRP	To encourage owners and operators who continue to maintain and use land that is currently in the CRP in an environmentally sustainable way. Keeping these acreages in grass cover is intended to help preserve most of the environmental benefits established under general CRP.[a]	0 to 20
Existing CRP cover	To provide additional points depending on the current CRP cover.	0 or 10
Beginning, limited resource, socially disadvantaged, or veteran farmers and ranchers	To encourage landowners and operators who meet certain demographic requirements, set by USDA, to submit offers for their land.	0 or 10
Maximizing grassland preservation	To encourage continued grassland preservation in areas subject to threat of conversion.	0 to 35
Land under threat of conversion (county)	To assign up to 10 points if over 50 percent of the offered acreage is in a county under threat of conversion.	0 to 10

Factors and subfactors	Purpose	Possible points
Land under threat of conversion	To assign up to 20 points if over 50 percent of the offered acreage is within an area identified as under a national threat of conversion.	0 or 20
Contract length	To assign 5 points if the offer is for a 15-year contract.	0 or 5
Vegetation cover	To assign points for practices on the land offered.	0 to 30
Environmental factors	To encourage conservation in areas of designated candidate, threatened, or endangered species or critical habitat.	0 to 45
Wildlife priority zone	To assign 15 points if over 50 percent of the offered acreage is within an approved state wildlife priority zone.	0 or 15
State grassland priority enrollment criteria	To assign 15 points if over 50 percent of the offered acreage is in a state grassland CRP zone.	0 or 15
Grassland CRP national priority zone	To encourage continued grassland preservation and to provide an incentive for environmentally sensitive areas. If over 50 percent of the offer is in one of the national priority zones, the offer will receive 15 points.	0 or 15
Small-scale livestock operation	To assign points if the landowner or operator certifies that they are eligible for a small-scale livestock operation, which is an operation with 140 grazing animal units or fewer.b Landowners also certify that they will not offer more than 200 acres for the program.	0 or 10
Cost factor	To consider cost as a factor in selecting offers. The subfactor used is dependent on the location of the land.	0 to 25
Cost subfactor	To assign points to offers for land located in counties in which the maximum rental rate is less than or equal to $15 per acre.	0 or 15
Offer less than maximum rental rate	To assign points to offers with a rate that is lower than the respective county maximum rental rate.	0 to 25

Source: GAO analysis of U.S. Department of Agriculture (USDA) information. | GAO-24-106311.

[a] Operators include those who lease the land or otherwise have the landowner's approval to enroll in the program.

[b] Different kinds of animals are assigned different units, per the Farm Service Agency's handbook for CRP. For example, a dairy cow is 1.40, a heifer is 0.86, and a goat is 0.15.

Appendix XI: Comments from the U.S. Department of Agriculture

USDA United States Department of Agriculture
Farm Production and Conservation
Farm Service Agency
1400 Independence Avenue, SW
Mail Stop 0510
Washington, DC 20250

DATE: **August 20, 2024**

TO: **Steve Morris**
Director, Natural Resources & Environment
General Accountability Office

FROM: **Steven Peterson** STEVEN PETERSON Digitally signed by STEVEN PETERSON Date: 2024.08.22 08:16:45 -04'00'
Associate Administrator

THROUGH: **John J. Berge**
***Acting* Deputy Administrator for Farm Programs**

SUBJECT: **Government Accountability Office (GAO) Audit # GAO-24-106311—"Improving How USDA Selects Land Could Increase Environmental Benefits" Draft Report for Recommendations, 1, 2, 3 and 4**

The U.S. Department of Agriculture (USDA), Farm Service Agency (FSA) is submitting responses for which FSA is responsible be accepted based on the support of the information provided in this memorandum. FSA is submitting four recommendations for executive action to GAO.

RECOMMENDATION 1:

The Administrator of USDA's Farm Service Agency should build on existing efforts by incorporating, in a timely manner and as appropriate, reliable science-based information from models or other sources into the agency's methods for assigning points that are used to select offers for the Conservation Reserve Program.

Agency Response

USDA agrees that utilizing recent studies and models provide a more reliable, science-based approach for assigning points used to rank and select offers for the competitive general CRP and Grassland CRP signup types. Extensive updates to the current scoring methods may require significant resources, analysis of impacts to the program, and may take considerable time. Resources and factors to consider are software platform, policy requirements, testing, and implementation. Implementation will be impacted by any new legislation such as farm bill and it may affect resources and factors mentioned. Funding may be necessary and timeframes for completion may need extension. It is also noted that models used must be readily available and updated to support Nationwide implementation of the CRP.

RECOMMENDATION 2:
The Administrator of USDA's Farm Service Agency should develop a process to periodically review its methods for selecting offers for the Conservation Reserve Program. Such a process should specify the frequency of the reviews and include soliciting and publicly reporting on suggestions from experts, including non-USDA scientists and economists.

Agency Response
USDA agrees with the recommendation to periodically review its methods for selecting CRP offers. Although a process is still underway and due to the complexity of the methods and the time constraints associated with changes, USDA believes a thorough review should be conducted every 5 years. This aligns with the current farm bill cycle and provides adequate time to solicit comments, review the comments and determine whether updates and changes should be made relative to the comments received. USDA will need to consider how publicly reporting suggestions from experts, including non-USDA scientists and economists will be meaningful and incorporated to the implementation of CRP.

RECOMMENDATION 3:
The Administrator of USDA's Farm Service Agency should address any findings resulting from its periodic reviews to enhance its methods for selecting offers for the Conservation Reserve Program, as appropriate.

Agency Response
USDA agrees to address any findings that are a result of the periodic reviews to enhance its methods for selecting offers for CRP.

RECOMMENDATION 4:
The Administrator of USDA's Farm Service Agency should annually analyze information on environmental benefits by type of enrollment and for the entire program, and publicly report the information, such as on its website.

Agency Response
USDA agrees to develop strategies for an annual analyzation on information regarding environmental benefits by type of enrollment and for the entire program while publicly reporting information.

2

Appendix XII: GAO Contact and Staff Acknowledgments

GAO Contact

Steve D. Morris, (202) 512-3841 or morriss@gao.gov

Staff Acknowledgments

In addition to the contact named above, Nkenge Gibson (Assistant Director), John Barrett (Analyst in Charge), Xiang Bi, Kevin Bray, Gary Brown, Tara Congdon, Thomas Cook, Cindy Gilbert, Benjamin T. Licht, Grant Mallie, Jason Rodriguez, Alec McQuilkin, Cynthia Norris, Beverly Peterson, Dan Royer, and Wes Sholtes made significant contributions to this report.

GAO's Mission

The Government Accountability Office, the audit, evaluation, and investigative arm of Congress, exists to support Congress in meeting its constitutional responsibilities and to help improve the performance and accountability of the federal government for the American people. GAO examines the use of public funds; evaluates federal programs and policies; and provides analyses, recommendations, and other assistance to help Congress make informed oversight, policy, and funding decisions. GAO's commitment to good government is reflected in its core values of accountability, integrity, and reliability.

Obtaining Copies of GAO Reports and Testimony

The fastest and easiest way to obtain copies of GAO documents at no cost is through our website. Each weekday afternoon, GAO posts on its website newly released reports, testimony, and correspondence. You can also subscribe to GAO's email updates to receive notification of newly posted products.

Order by Phone

The price of each GAO publication reflects GAO's actual cost of production and distribution and depends on the number of pages in the publication and whether the publication is printed in color or black and white. Pricing and ordering information is posted on GAO's website, https://www.gao.gov/ordering.htm.

Place orders by calling (202) 512-6000, toll free (866) 801-7077, or TDD (202) 512-2537.

Orders may be paid for using American Express, Discover Card, MasterCard, Visa, check, or money order. Call for additional information.

Connect with GAO

Connect with GAO on Facebook, Flickr, Twitter, and YouTube. Subscribe to our RSS Feeds or Email Updates. Listen to our Podcasts. Visit GAO on the web at https://www.gao.gov.

To Report Fraud, Waste, and Abuse in Federal Programs

Contact FraudNet:

Website: https://www.gao.gov/about/what-gao-does/fraudnet Automated answering system: (800) 424-5454 or (202) 512-7700

Congressional Relations

A. Nicole Clowers, Managing Director, ClowersA@gao.gov, (202) 512-4400, U.S. Government Accountability Office, 441 G Street NW, Room 7125, Washington, DC 20548

Public Affairs

Sarah Kaczmarek, Acting Managing Director, KaczmarekS@gao.gov, (202) 512- 4800, U.S. Government Accountability Office, 441 G Street NW, Room 7149 Washington, DC 20548

Strategic Planning and External Liaison

Stephen J. Sanford, Managing Director, spel@gao.gov, (202) 512-4707

U.S. Government Accountability Office, 441 G Street NW, Room 7814, Washington, DC 20548.

Chapter 9

Land Use of Rejected, Enrolled, and Expiring Fields in the Conservation Reserve Program*

Andrew Rosenberg
Bryan Pratt
David Arnold
and Ryan Williams

Cover photo from USDA Flickr.

Recommended Citation Format for This Publication

Rosenberg, A., Pratt, B., Arnold, D., & Williams, R. (2024). Land use of rejected, enrolled, and expiring fields in the Conservation Reserve Program

---

* This is an edited, reformatted and augmented version of U.S. Department of Agrticulture Economic Research Service, Economic Information Bulletin Publication No. 276, dated August 2024.

In: The Conservation Reserve Program
Editor: Jerry P. Norton
ISBN: 979-8-89530-774-8

(Report No. EIB-276). U.S. Department of Agriculture, Economic Research Service.

Use of commercial and trade names does not imply approval or constitute endorsement by USDA.

To ensure the quality of its research reports and satisfy governmentwide standards, ERS requires that all research reports with substantively new material be reviewed by qualified technical research peers. This technical peer review process, coordinated by ERS' Peer Review Coordinating Council, allows experts who possess the technical background, perspective, and expertise to provide an objective and meaningful assessment of the output's substantive content and clarity of communication during the publication's review.

In accordance with Federal civil rights law and U.S. Department of Agriculture (USDA) civil rights regulations and policies, the USDA, its Agencies, offices, and employees, and institutions participating in or administering USDA programs are prohibited from discriminating based on race, color, national origin, religion, sex, gender identity (including gender expression), sexual orientation, disability, age, marital status, family/parental status, income derived from a public assistance program, political beliefs, or reprisal or retaliation for prior civil rights activity, in any program or activity conducted or funded by USDA (not all bases apply to all programs). Remedies and complaint filing deadlines vary by program or incident.

Persons with disabilities who require alternative means of communication for program information (e.g., Braille, large print, audiotape, American Sign Language, etc.) should contact the responsible Agency or USDA's TARGET Center at (202) 720-2600 (voice and TTY) or contact USDA through the Federal Relay Service at (800) 877-8339. Additionally, program information may be made available in languages other than English.

To file a program discrimination complaint, complete the USDA Program Discrimination Complaint Form, AD-3027, found online at How to File a Program Discrimination Complaint and at any USDA office or write a letter addressed to USDA and provide in the letter all of the information requested in the form. To request a copy of the complaint form, call (866) 632-9992. Submit your completed form or letter to USDA by: (1) mail: U.S. Department of Agriculture, Office of the Assistant Secretary for Civil Rights, 1400 Independence Avenue, SW, Washington, D.C. 20250-9410; (2) fax: (202) 690-7442; or (3) email: program. intake@usda.gov.

USDA is an equal opportunity provider, employer, and lender.

Abstract

The U.S. Department of Agriculture's Conservation Reserve Program (CRP)—the agency's largest land retirement program—pays landowners a yearly rental fee to plant and maintain environmentally beneficial land covers on eligible portions of their land instead of crops. In this report, the authors examine land-use outcomes of parcels that were both offered and rejected from General Signup 49, the auction component of CRP that was administered in 2016. Because Signup 49 featured an unusually high rejection rate (81.6 percent), many offers rejected from the auction in this CRP Signup would likely have been accepted in more typical Signups. Examining land-use choices of these rejected offers helps iden- tify what land uses would have been replaced by CRP land covers if the Signup had featured a higher acceptance rate. The study finds that rejected CRP land goes into a variety of land covers, including cropland, grassland, and Continuous Signup CRP (allowing environmentally sensitive land to be enrolled at any time). The study also finds that new program applicants who are rejected are more likely to be in cropland after the Signup, if not in the CRP, compared to returning participants. Finally, the study finds large geographic differences in land-use decisions of rejected CRP applicants and discusses how the current design of the CRP influences its impacts.

Keywords: carbon sequestration, conservation covers, conservation practices, conservation program, Conservation Reserve Program, Environmental Benefits Index, land retirement, land use, rejected offers, soil erosion, wildlife habitat

Acknowledgments

The authors thank Catherine Feather, Rich Iovanna and other USDA, Farm Production and Conservation, Business Center staff for assistance with data and valuable insights drawn from considerable experience with the Conservation Reserve Program. We also thank staff of the Office of Energy and Environmental Policy in USDA's Office of the Chief Economist; and Megan Stubbs of the Congressional Research Service, for helpful insights. The authors also thank several anonymous reviewers and ERS economists for helpful technical feedback. In addition, the authors thank the ERS Publishing Services Branch (PSB) for editorial services.

About the Authors

Andrew Rosenberg and Bryan Pratt are economists with the USDA, Economic Research Service. David Arnold is a geographer with the USDA, Economic Research Service. Ryan Williams is a geographer with the USDA, Natural Resources Conservation Service.

What Is the Issue?

The U.S. Department of Agriculture's Conservation Reserve Program (CRP), administered by the USDA, Farm Service Agency, pays landowners to plant and maintain environmentally beneficial land covers instead of crops. In this report, the authors determined the extent to which CRP removes intensive land uses from production, examining land-use decisions made by the owners of parcels offered and rejected from the 2016 CRP General Signup. The 2016 CRP General Signup had a particularly high rejection rate for a General Signup, the CRP's competitive auction. The high rejection rate means that many rejected offers from the 2016 Signup would likely have been accepted in more typical Signups, and the observed land-use choices of rejected offers from 2016 serve as an estimate of what land uses are replaced when land is enrolled in CRP in other Signups. The authors also examined the tradeoffs that would be associated with prioritizing new or returning CRP applicants or with maximizing total acreage or the proportion of cropland enrolled in contracts.

What Did the Study Find?

The study compared land-use transitions of accepted and rejected offers from the 2016 General Signup with the following results:

- After rejection, 16.6 percent of acres that landowners offered were planted in corn and soybeans, 23.4 percent were in wheat or left fallow, 20.7 percent in grassland, 14.7 percent in mixed forage, and 8.9 percent were idled or were in other land uses. Landowners enrolled the remaining 15.7 percent of offered and rejected acres in the Continuous Signup CRP (an alternative CRP enrollment mechanism which allows environmentally sensitive land to be enrolled throughout the year).

- Rejected offers with higher scores for the Environmental Benefits Index (EBI), the General Signup CRP's multicategory score used for ranking and selecting offers (see definition box), had lower proportions of land in corn and soybeans and higher proportions in wheat.

The authors also estimated the average acreage in each land use that could be retired for every additional $100 spent enrolling land in CRP. They found that:

- If rejected offers that did not subsequently enter CRP through Continuous Signup were instead accepted, the program would retire a total of 1.39 acres on average per $100 spent, including: 0.28 acres of corn and soybeans, 0.23 acres of wheat, 0.12 acres of other cropland, 0.24 acres of mixed forage, 0.33 acres of grassland, 0.01 acres of timberland, and 0.18 acres of fallow or idle land;
- The total acreage in cropland retired per $100 spent would decline if the EBI threshold was set lower, as in other Signups. However, acres in corn and soybeans retired per $100 spent would increase with a decreasing threshold.

Land-use outcomes of rejected offers vary with recent CRP participation prior to applying in 2016:

- Many returning applicants, those with recently expired CRP contracts, reentered the CRP through Continuous Signup after being rejected from General Signup. Returning applicants were more likely than new applicants to have land in grassland and mixed forage after their offers were rejected. New applicants were much more likely to grow crops after their offers were rejected;
- Enrolling new applicants would retire more cropland per dollar spent than reenrolling returning applicants but would also result in less total acreage enrolled per dollar spent.

The report also compared fields that landowners removed from the CRP voluntarily and involuntarily with the following results:

- Landowners that attempted to reenroll in the General Signup CRP but were rejected ("involuntary exits") made similar land-use decisions to those that exited the CRP without attempting to reenroll ("voluntary exits"). However, involuntary exits were less likely to lead to crop planting than voluntary exits after exiting the CRP.

Land-use outcomes of rejected offers also differed geographically. The authors found that:

- Corn Belt States had the highest proportions of non-CRP land with crops. Western States had more land in grassland and mixed forage. Several States had close to half of their land that was rejected from General Signup entered into the Continuous Signup CRP.
- Acreage potentially retired per $100 varied considerably by State. A high amount of acreage per $100 could be retired in Western States due to their low rent land. Corn Belt States could retire the most acres in corn and soybeans, but States with lower rental rates could retire more cropland acres per dollar spent.

How Was the Study Conducted?

The authors observed field-level land-use decisions of all offers made in the 2016 General Signup. They utilized offer data on acceptance status, offered rental rates, practices offered, and EBI scores. Using the USDA, Farm Service Agency's (FSA) Common Land Unit dataset, they linked most offers to land-use data recorded in FSA's Crop Acreage Reporting Database. Land-use information from 2013 through 2019 was pooled into two categories: the 4 years prior to the 2016 Signup (2013–16) and the 3 years after (2017–19).

Definitions

Accepted offers – Offers submitted to and accepted into the 2016 General Signup. When discussing accepted offers throughout this report, we exclude offers that were accepted but withdrawn. We also do not include offers that were initially rejected but ultimately secured a contract.

Enrolled offers – Offers submitted to the 2016 General Signup that then were enrolled in a contract through that Signup, regardless of whether the offer was initially accepted or rejected. Some offers are initially rejected but later accepted upon appeal and then enrolled in the program.

Environmental Benefits Index – The Environmental Benefits Index (EBI) is used to determine the relative environmental benefits of parcels being offered into the General Signup. The EBI score is made up of six components. The first five components reflect benefits to wildlife habitat, water quality, erosion, enduring benefits beyond the contract period, and air quality, respectively. The final component reflects the cost of enrolling the land being offered to CRP.

Environmental EBI – The Environmental EBI is the Environmental Benefits Index (EBI) without the cost component, including both points for environmental sensitivity of the land and points for the environmental benefits of the practice choice.

Expiring applicants – Offers submitted to the 2016 General Signup with a CRP contract expiring at the end of 2016, so that enrollment in the 2016 General Signup would lead to a new contract beginning the day after the then-current contract expired.

Involuntary exits – Fields that leave CRP at the end of the contract and do not reenter the General Signup CRP, despite the landowner applying for reenrollment. The fields also are not reenrolled into the CRP via the Continuous Signup during our sample period. Such fields are involuntarily leaving the program at the fields' designated departure date. These fields—comprising rejected offers that were offered for reenrollment (returning applicants)—are included in "Rejected offers."

New applicants – Offers submitted to the 2016 General Signup with no history of CRP enrollment during 2013 through 2016.

Non-offers – Fields that were not offered for enrollment.

Rejected offers – Offers submitted to and rejected from the 2016 General Signup. When discussing rejected offers throughout this report, we exclude those offers that were rejected initially but ultimately secured a contract. We also do not include offers that were accepted but withdrawn.

Returning applicants – Offers submitted to the 2016 General Signup, with CRP enrollment during at least 1 year during the period 2013 through 2016.

Unenrolled offers – Offers submitted to the 2016 General Signup that were not enrolled in a contract through that Signup, regardless of whether the offer was initially accepted or rejected. Some offers are initially accepted but were withdrawn during the enrollment process and these offers are never enrolled in a contract through this Signup.

Voluntary exits – Fields that leave the CRP at the end of the contract (or before) without the attempt to reenroll in the General Signup CRP. These fields also do not reenter CRP during the study's sample period via Continuous Signup. Such fields are voluntarily leaving the program at (or before) their designated departure. Voluntary exits are a subset of nonoffers, defined above.

Introduction

The U.S. Department of Agriculture (USDA) spends nearly $2 billion annually on the Conservation Reserve Program (CRP), administered by USDA's Farm Service Agency (FSA). Most of the funding goes to landowners and operators to retire cropland for 10- to 15-year periods (USDA, FSA, 2022). The program aims to reduce soil erosion, increase wildlife habitat, improve water and air quality, and sequester carbon, chiefly by removing environmentally sensitive land from agricultural production and establishing land covers that advance environmental goals. This report measures the degree to which the CRP displaces intensive land uses from production and explores the relationship between the program's current enrollment mechanism and land-use outcomes that influence the program's environmental goals.

This report assesses the land-use impacts of the CRP, mostly by examining land-use decisions of landowners and operators whose offers were rejected from the 2016 General Signup. The General Signup enrollment mechanism has a competitive format that ranks offers on environmental benefits and cost to enroll, with an offer accepted if it scores above a threshold determined after submissions close. Rejected offers are unique in that we can observe both what landowners and operators[81] actually do with their land after their offers are rejected, as well as what they would have done if their offers were instead accepted. If the offers had been accepted, landowners would have been contractually obligated to plant and maintain the conservation covers in their offer. The report assesses land-use decisions of landowners with rejected offers from a particularly stringent auction, General Signup 49, conducted in 2016. Post-Signup land use decisions on parcels rejected from Signup 49 provide a relevant estimate of CRP's land-use impacts for a large portion of offered land in most Signups, since more than 80 percent of the rejected offers from the 2016 Signup would have been accepted in most other General

[81] Landowners and operators are often referenced as just "landowners" throughout the report.

Signups.[82] The authors also estimate the costs of avoiding additional intensive land uses through CRP, linking proposed CRP rental rates with subsequent land-use decisions.

The report begins with a broad analysis of landowner land-use decisions in the years immediately following Signup 49 on land that was rejected from the 2016 Signup. In the 3 years afterward, 37.7 percent of acreage rejected from the Signup went into crop production, including 13.7 percent in wheat, 8.9 percent in soybeans, and 7.7 percent in corn. An additional 9.7 percent of acreage was fallowed, typically in a wheat-fallow rotation. After crop production, the next most common land use was grass at 20.7 percent of acreage, usually for grazing. An additional 14.7 percent of acreage was in mixed forage. Less than 1 percent of acreage was either idled or used in timber production. Notably, landowners rejected from the General Signup enrolled 15.7 percent of their acreage into CRP through the Continuous Signup, generally at a higher cost to FSA. These results were broadly similar when looking only at rejected acreage that would have been accepted in the four previous General Signups. The authors find that land-use choices of landowners that made rejected offers were relatively stable across Environmental Benefits Index (EBI) values (see the Definitions box entry for EBI), by share of acreage, with a gradual increase in commodity crop production as EBI scores fell.

Next, the authors examine how land-use impacts of CRP vary for some key subpopulations of offers, relevant to two important policy tradeoffs related to land-use impacts of the CRP General Signup. The first tradeoff is whether to prioritize new applicants or returning applicants for enrollment in CRP. The authors find that new applicants would be much more likely than returning applicants to grow crops when not in a CRP contract. More than 60 percent of acreage from new applicants was planted to crops in the first 3 years after rejection from the 2016 General Signup, that is, among rejected offers not going into Continuous Signup CRP after rejection. In contrast, less than 25 percent of rejected acreage from returning applicants was subsequently planted to crops among those not entering CRP through Continuous Signup. The most common planting of landowners seeking reenrollment was grass, followed by mixed forage. In addition, a large minority of the landowners seeking reenrollment were able to secure a Continuous Signup contract, which

[82] The 2016 General Signup had a very low acceptance rate, as detailed in table 1, because the program's enrollment limit had recently been reduced by Congress. The enrollment constraint has since risen.

generally pays higher rates than General Signup contracts. Enrolling new applicants would lead to greater cropland retirement per dollar spent in the program. However, there may be environmental benefits for which retaining land in conservation cover would take priority, despite the lower amount of intensive land being displaced. Currently, the EBI provides no benefit or penalty for prior enrollment status.

To further investigate the consequences of landowners with expiring land exiting the program, the authors examine the extent to which landowners voluntarily and involuntarily exiting the program differ in subsequent land use. Landowners are considered to be voluntarily exiting CRP if they possessed an expiring CRP contract in 2016 but did not submit an offer, while a landowner is considered to involuntarily exit CRP if they possessed an expiring CRP contract and did submit an offer to the 2016 General Signup. While landowners voluntarily exiting CRP were more likely to grow crops than those involuntarily exiting CRP, the differences were not large, and cropping occurred on a minority of acres across both groups. This report only includes 3 years of data after program exit, but this finding suggests that both voluntarily and involuntarily exiting landowners often persist in lower intensity land uses, such as grassland for grazing or forage.

The second tradeoff is whether to maximize either total acreage enrolled in CRP (including all land uses retired) or the proportion of acres in contracts representing retirement of intensively managed cropland. The EBI rewards offers with lower rental rates, leading to an intended decline in the number of acres retired per dollar spent by the program as EBI scores decrease. However, regions with higher rental rates also tended to have a higher amount of acres growing valuable commodity crops, which we found leads to a small increase in the share of acreage in corn and soybeans as EBI scores decrease. In the context of this second tradeoff, it is a policy challenge to decide how the program (which is constrained by the maximum number of acres that can be enrolled) balances the goals of retiring the most acreage per dollar spent while also enrolling intensively used land, all the while targeting land with the highest environmental value. This report demonstrates that prioritizing land in terms of this tradeoff relates strongly with decisions about where land is enrolled, showing the considerable geographic differences of land-use impacts of CRP, as well as the costs of retirements per acre.

Estimates of land-use impacts of the CRP have played an important role in many previous evaluations of the program. Hansen (2007) identified three main approaches that have been used in the literature to estimate these impacts. The first approach that uses actual land-use outcomes assumes that CRP lands would remain in the same uses they were in prior to enrolling in the CRP if landowners had not enrolled their land. For example, in estimating water quality impacts of the CRP, Ribaudo (1989) assumed that soil erosion levels would remain constant had the CRP not existed, implicitly assuming that land use would have remained constant as well. This first approach was only feasible for all enrolled CRP land for the initial CRP General Signup. Given that a very large proportion of CRP offers are returning applicants (61 percent of acres in 2016) and that new and returning applicants are very different, this approach is no longer feasible (as we show in this report).

The second major approach that uses observed land-use outcomes, outlined in Hansen (2007), assesses CRP land-use impacts with expiring contracts (e.g., Roberts & Lubowski, 2007; Hendricks & Er, 2018; Bigelow et al., 2020; Morefield et al., 2016). Some studies have used this approach primarily to evaluate the specific impacts of exiting CRP land (e.g., Bigelow et al., 2020), while others have used CRP exits to infer overall landscape impacts of CRP (Roberts & Lubowski, 2007).

The third main approach outlined in Hansen (2007) is to use survey data on CRP participants' intended uses if the CRP didn't exist. For example, in 1993, the Soil and Water Conservation Society conducted a national survey of farmers with land enrolled in CRP. Their study found that farmers would return about 63 percent of their land to cropland if contracts were not renewed (Dodson et al., 1994). Barnes et al. (2020) used survey data to detect selection effects among farmers previously enrolled in CRP in the Western Plains. The authors of the study asked farmers with land previously in CRP how they were currently using their land. The authors compared responses from those farmers who applied to reenroll (but were rejected) with those who did not apply for reenrollment. Those farmers who applied to reenroll were about 14 percentage points less likely to convert their land to crops after their contracts expired. The authors of that article also found a similar difference between farmers currently in the CRP who expect to reenroll or who do not.

The approach taken in the current report expands upon these previous approaches. Landowners with exiting CRP land may be significantly different than landowners that are new to CRP, especially if former participants are slow to change back to intensive uses after the exit. As found by Hendricks and Er (2018), land that transitioned into the CRP was more likely to have come from cropland than land that transitioned out of the CRP was to enter into cropland. The authors hypothesized that the difference may be due to either high costs of land conversion or to nonadditionality, occurring if landowners would have put their land into less productive uses in the absence of the program. Roberts and Lubowski (2007) argued that the low rate of return to cropland among CRP exits was more likely to come from conversion costs than nonadditionality, and that even a small land conversion cost could lead to significant persistence. In the current report, the authors evaluate land-use impacts from both populations—new applicants as well as returning applicants—and examine the differences between land-use impacts of these groups. The findings in the current report expand upon the findings of Rosenberg and Pratt (2024), who estimated land-use impacts of CRP using a regression discontinuity design, based on the EBI threshold. The current report also elaborates on the findings in Barnes et al. (2020), further investigating the differences in land-use outcomes for voluntary and involuntary CRP exits.

The current study also examines how effective the CRP was at obtaining land-use impacts per dollar spent, which is important for examining the cost-effectiveness of the program. Several previous studies have examined other aspects of cost-effectiveness for the CRP (Kirwan et al., 2005; Miao et al., 2016; Hellerstein et al., 2015; Cramton et al., 2021; Smith, 1995). Kirwan et al. (2005) examined CRP offers to estimate the premiums paid to enroll land, measured as the difference between rental rates paid and reservation rents.

Further, in a study closely related to our study, Miao et al. (2016) (also using CRP offer data) assessed the cost-effectiveness of the EBI at obtaining environmental benefits using components of the EBI to measure benefits. However, to our knowledge, no existing studies incorporate observed land-use impacts of CRP into evaluations of cost-effectiveness.

CRP Background and History of the General Signup Mechanism

The CRP is USDA's primary agricultural land retirement program. Administered by USDA's Farm Service Agency (FSA), the program primarily provides annual rental payments to landowners and operators to establish conservation cover on land with a history of crop cultivation, as well as partial reimbursement for the costs of establishing such cover. The purpose of the program is to improve air and water quality, wildlife habitat, and carbon sequestration. The mechanism by which the program generates these benefits is through land-use change—specifically, the conversion of intensive crop acreage into perennial cover for 10–15 years, particularly on highly erodible or environmentally important land.

The program comprises three subprograms: General Signup, Continuous Signup, and Grasslands Signup.[83] The General Signup is the CRP subprogram with a competitive auction enrollment mechanism for cropland. The General Signup solicits offers from landowners to enroll in CRP. These offers are ranked according to the Environmental Benefits Index (EBI) and FSA rejects offers below a threshold determined after all offers are submitted.[84] In addition to the national EBI threshold, some offers are rejected due to additional constraints on the program, requiring CRP enrollment to comprise no more than 25 percent of all cropland in any U.S. County. For offers in counties where this restriction is relevant, offers are accepted starting with the highest EBI scores and ending with the lowest EBI scores until the cap is met or the national cutoff is reached.[85] The mechanism of the General Signup permits the analysis presented in this report, as it is possible to observe the actual behavior of fields with legally binding offers to establish conservation cover at an agreed rental rate. For fields associated with rejected offers, it is possible to observe their non-CRP land-use trajectory.

To be eligible for the General Signup, the land being offered must be either highly erodible (greater than or equal to a set threshold) or located within a designated Conservation Priority Area (CPA); and the land must have a sufficient cropping history.[86] Alternatively, land that is in the final year of a CRP contract is also eligible to be reenrolled.[87] EBI scores are awarded points based on estimates of the environmental sensitivity of offered land, as well as the environmental benefits of the cover proposed and the rental rate requested. If parcels are accepted into General Signup, the parcels are typically in a 10-year contract.

The Continuous Signup has become increasingly important in recent years. With few exceptions, enrollment in the Continuous Signup is first-

---

[83] The Continuous Signup includes the Conservation Reserve Enhancement Program.

[84] The 2018 Agricultural Improvement Act expanded the CRP to include a Grasslands subprogram, which also uses a competitive auction enroll- ment mechanism. However, the Grasslands subprogram enrolls not existing cropland but existing grasslands (including grazing land) deemed likely to be converted to cropland in the absence of the program. Extending this analysis to the Grasslands subprogram is beyond the scope of this report.

[85] The countywide restriction of no more than 25 percent of cropland acres in CRP applies to the sum of all CRP subprograms. Offers are rejected if their acreage would lead the total county enrollment in CRP the following October to exceed the allowed acreage.

[86] Specifically, the candidate parcel must have been cropped during 4 of 6 years prior to enrollment, as specified in the Farm Bill. For the 2016 General Signup, offered land must have been cropped 4 of the years from 2008 through 2013. For the purposes of eligibility as cropland (7 CFR 1410.6(b)(1)), land that is enrolled in CRP is 'considered planted' as cropland (7 CFR 1410.2 "Considered planted").

[87] 16 U.S. Code \S 3831, d, 1.

come, first-served. The Continuous Signup involves stricter eligibility requirements than the General Signup, and in return, the Continuous Signup offers higher incentives. In addition to annual rental rates similar to those paid in the General Signup, landowners that enroll in the Continuous Signup may also be awarded additional annual or one-time incentive payments. Many land- owners opt to enroll in the Continuous Signup after being rejected from the General Signup, suggesting that at least some landowners prefer the General Signup. However, we do not attempt to assess the choice between the two enrollment mechanisms in this report. The Conservation Reserve Enhancement Program (CREP) constitutes a significant portion of the Continuous Signup. Projects with CREP involve funding from USDA and U.S. State partners. The projects identify high-priority resource concerns and provide incentives to farmers in specified regions that can exceed those from other CRP programs (USDA, FSA, 2021).

The 2016 General Signup, CRP Signup 49, is a uniquely useful Signup for multiple reasons in understanding the land use of fields rejected from the CRP. First, the 2016 General Signup had the highest rate of rejection in the history of the contemporary CRP General Signup at approximately 82 percent (table 1). The result is a large sample of fields with rejected offers—specifically, 33,629 fields associated with a rejected offer that also have administrative land-use records. Second, the 2016 General Signup occurred in a period of legislated reductions to the size of the program. As a consequence, these fields had no other opportunity for enrollment through a General Signup from 2014 through 2019. As a result, most land associated with rejected offers subsequently went into non-CRP land uses in the 3 years following the 2016 Signup unless the land- owners were able to enroll through the Continuous Signup. Given the landowners' intentions to join the CRP through the General Signup in 2016, they may have entered the program through the Continuous Signup as soon as 2017.

Table 1. USDA Conservation Reserve Program (CRP) general signup offer acceptance and rejection rates, 1997–2021

Signup	1997 (15)	1997 (16)	1998 (18)	1999 (20)	2003 (26)	2004 (29)	2006 (33)	2010 (39)	2011 (41)	2012 (43)	2013 (45)	2016 (49)	2020 (54)	2021 (56)
Accepted	63.8	59.6	68.2	70.4	54.3	75.7	79.4	91.6	77.2	87.7	87.0	18.4	90.9	88.3
Rejected	36.2	40.4	31.8	29.6	45.7	24.3	20.6	8.4	22.8	12.3	13.0	81.6	9.1	11.7
County cap	1.17	1.89	2.21	1.72	1.03	1.01	1.57	2.40	2.40	1.28	0.82	0.18	0.43	1.47
Low EBI	35.1	38.5	29.6	27.9	44.6	23.3	19.0	6.05	20.4	11.1	12.2	81.4	8.68	10.2
EBI cutoff	259*	247*	245*	246*	269	248	242	200	221	209	209	292	210	175^
Median EBI	282`	262`	267`	270`	274`	274`	273`	273	262	265	254	249	274`	235`
Offers (thousands)	252	126	90	56	71	26	23	50	39	48	28	26	57	23
Rental rate index	51.3	51.3	51.4	52.1	56	58.2	63.5	92.7	100	101.4	105.8	130.4	124.5	124.5
Rental rate points for a typical field	133.2	87.6	87.5	87.0	88.6	87.2	87.6	74.3	70.3	69.6	67.2	53.7	62.6	62.6

Source: USDA, Economic Research Service using Conservation Reserve Program (CRP) General Signup offer data from USDA, Farm Service Agency and rental rate index data from USDA, National Agricultural Statistics Service.

EBI= Environmental Benefits Index.

Note: In the first four rows, the number in each cell represents the percentage of offers in a given category within a General Signup. The "County cap" and "Low EBI" categories are subsets of the "Rejected" category. The EBI cutoff row shows the minimum EBI score accepted at a national level. Those cutoffs marked with an asterisk (*) are from Signups 15–20, which had more total EBI points available than in 2016. The total EBI score for Signups 26 through 56 is 545. Signup 15 had a total possible EBI score of 600, and Signups 16–20 had total possible EBI scores of 560. The (^) mark for 2021 is to note that some U.S. States accepted offers with EBI at or above 165 based on legislative language on the distribution of CRP land across States. The Median EBI row provides the median

EBI score of all offers in a given Signup. Those values with a mark (`) are from Signups with at least slightly different EBI formulas relative to 2016. The Offers row presents the number of offers in that General Signup in thousands. The rental rate index is a national estimate of cash rents relative to 2011, the base year. The index is calculated as the national survey estimate for the average rental rate paid for nonirrigated cropland in a given year, divided by the survey estimate for 2011, multiplied by 100. Rental rate points for a typical field show the points awarded for a rental rate to a field with a rental rate equivalent to the median in 2016 ($96.21/acre) adjusted by the rental rate index. For example, the 2011 value represents the points awarded for a rental rate of $73.78, calculated from 96.21*100/130.4. The 2016 Signup had a lower value largely because of the high rental rate index and unchanging EBI parameters for rental rates from prior years. The Signup examined in this report is highlighted in blue with rejection rates highlighted in orange.

The 2016 General Signup is also useful because many offers were rejected that likely would have been accepted in other years. With its historically high EBI cutoff and rejection rate, the 2016 Signup had many offers that would have been accepted in 2013 or 2020, for example. It is likely that landowners would have anticipated the competitiveness of the 2016 Signup to some extent, given the accrual of demand for the program since 2013, which would be reflected in higher EBI scores. However, the median EBI score in 2016 was actually lower than in other Signups.[88] This finding is largely a result of inflation in rental rates. Adjusting for the changes in rental rates and the (lack of) changes in the EBI formula, the set of offers in 2016 had higher adjusted EBI scores than in 2012 and 2013 but lower than in 2020. Appendix figure B.1 illustrates the EBI distribution for General Signups in 2010 through 2020. Consequently, we might reasonably consider these rejected offers to be similar in many ways to offers that have historically been accepted into the program.

As a final note on these submitted offers, figure 1 shows the EBI distribution of 2016 General Signup offers, as well as whether offers ended up enrolled in CRP or not. Several points are notable in this distribution.

Figure 1 illustrates that the cutoff is strongly binding with limited exceptions. The primary exception was the rejection of offers in counties where the CRP acreage was at or near 25 percent of county cropland. As shown in table 1, only a small portion (0.18 percent) of total offers fall into this category. The other two exceptions were appeals and offers that failed to advance to enrollment. Some accepted offers failed to pass through subsequent steps leading to enrollment, while some rejected offers appealed for inclusion.[89] These processes were rare, with the exception that more than one-half of offers rejected due to county cap constraints were able to secure a contract. Less than 0.5 percent of offers with an EBI below the national cutoff were able to secure a 2016 General Signup contract. Just under 6 percent of accepted offers failed to ultimately enroll in a 2016 General Signup contract.

---

[88] Decomposing the EBI, 2016 offers did exhibit a higher median point total from cover practice choices.

[89] These formal processes are intended primarily to correct errors in the original offer that led to rejection and to catch offers omitted from the process in error. With respect to accepted offers not enrolling, all landowners and producers are allowed to withdraw their accepted offer within 15 days of notification. Offers may also be deemed infeasible after acceptance and such offers may be revised or withdrawn.

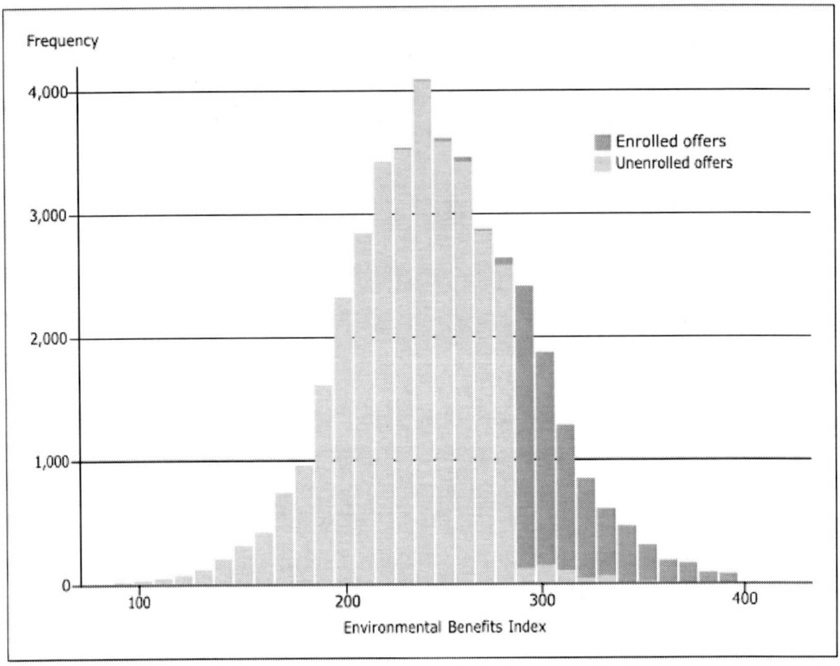

Source: USDA, Economic Research Service using Conservation Reserve Program offer and contracts data from USDA, Farm Service Agency.

Note: This figure is a histogram of fields by offered Environmental Benefits Index (EBI) score. Enrolled offers include accepted offers that enrolled, as well as rejected offers that were able to enroll upon appeal. Unenrolled offers include rejected offers unable to enroll through Signup 49, as well as accepted offers that either chose not to enroll or otherwise failed to enroll through Signup 49. The national threshold for acceptance was an EBI score of 292, while some counties had higher thresholds to comply with legislation requiring CRP land not to exceed 25 percent of county cropland.

Figure 1. EBI distribution of the 2016 General Signup-offered fields.

Data Used in the Analysis

The authors completed the analysis with field-level land-use data for the entire contiguous United States. We used two main sources of data and linked them to each other and through time using a panel of field information. First, we used detailed data on all offers submitted to the 2016 General Signup. Second, we derived land-use measures from USDA, FSA's Crop Acreage Reporting

Database (CARD), which provides administrative records of land use at the field level each year.

To identify what fields with rejected offers would have done had they been accepted into CRP, we used offer data from USDA, FSA. We observed not only the acceptance or rejection of each offer but also various attributes, such as the maximum allowed rental rate and the offered rental rate, the practice or practices offered, and the scores received for each component of the EBI. The offer data also contained field identifiers that could be used to identify the actual land-use decisions of the applicants after the offers were rejected. There were 26,279 offers made in General Signup 49. About 93 percent of the offers were geolocated, leaving 42,032 offered fields. Among these fields were 33,629 fields associated with rejected offers.

The main data to identify land-use decisions made by landowners after rejected CRP offers came from Form 578 data from USDA, FSA. All landowners are required to file a Form 578 on an annual basis in order to participate in USDA programs and receive financial support from USDA. The data generated through this process is compiled in the CARD, and this record serves as the official record of planted acreage for a variety of purposes. Reported plantings include a variety of useful details, including the crop planted, acreage planted to a specific crop, and the intended use of the planting. Each planting in CARD identifies the crop planted. This information can be used to identify CRP practices when a field is enrolled in CRP. Other than CRP land cover, the primary crops or land use categories used in this report are corn, soybeans, wheat, fallow or idle land, grassland, mixed forage, timberland, and other crops. Other crops include cotton, alfalfa, barley, sorghum and millet, among others that are found in small quantities. At times, corn and soybeans were combined in a single category based on the prevalence of corn-soy rotations, as were wheat and fallow because of the frequency of wheat-fallow rotations. In general, when we refer to crops or crop acreage in this report, we are referring to the combination of the land-use categories of corn, soybeans, wheat, and other crops. We refrain from using "cropland," as there is a specific meaning of this term in the context of the CRP.[90] When referring to all acres, we use terms such as total acreage.

[90] For the purposes of eligibility for CRP, cropland is defined by Title 7 of the Code of Federal Regulations. To be eligible for CRP, any field must be considered cropland or qualify under a specific alternative eligibility pathway. For this purpose, cropland is defined as land determined by a county committee to meet any one of seven conditions. These conditions generally cover land that is currently being used for crops for harvest or has been used for crops for harvest and is still capable of being used for crops for harvest, as well as a few

Fields were linked through time to allow for the observation of a panel of individual fields—associated with specific offers—from 2013 through 2019.[91] USDA, FSA's Common Land Unit (CLU) database is a continually updated spatial representation of all fields that interact with USDA, FSA. Because the database is designed to be accurate for the present, it can be challenging to trace a field through time as the field changes owner, operator, shape, or size. We spatially merged CLU polygons for 2016 through 2019 to identify 'fields,' which were the largest spatial unit with a common trajectory of CLUs over all years 2016 through 2019. For the purposes of this report, these spatial unions are referred to as fields, even when the unions comprise only a portion of what a landowner might describe as a field. Notably, data limitations mean that we could not observe land-use outcomes prior to a field-specific year between 2013 and 2016 for a small portion of fields. After limiting the sample to those offers that we could link to land-use data from before and after Signup 49, the analysis in this report used about 1.48 million offered acres (1.17 million of which were rejected) of the 1.86 million acres offered in Signup 49 (1.45 million of which were rejected).

The Land Uses of Rejected and Accepted Offers

This section begins with a broad overview of land-use outcomes before and after the 2016 Signup for land associated with both rejected and accepted offers. We then examine the land-use outcomes prior to and after the 2016 Signup for offers across the full range of EBI scores. This latter exercise helps to show that outcomes from rejected offers in 2016 provide insights about the land-use impacts of CRP from more typical Signups.

Figure 2 shows land-use transitions for rejected offers. The figure combines major crops into the categories discussed in the previous section, including corn-soybeans and wheat-fallow. We did this because, for the purposes of transition graphs, we needed to categorize fields into discrete categories in each multi-year period, and many crop fields are in rotations. Thus, we combined crops into common rotations to avoid having to identify a corn-soybeans field as either corn or soybeans in a given multi-year period,

---

additional unique circumstances (7 CFR 718). The panel ends in 2019 so it does not include the following General Signup in 2020. The reason is to minimize postrejected CRP land uses in the analysis.

[91] The panel ends in 2019 so that it does not include the following General Signup in 2020. This is to minimize postrejected CRP land uses in the analysis.

which may be misleading (in later figures, we depict the actual acreage in each disaggregated crop). The figure further includes the categories of grass and mixed forage, which are generally used for grazing and forage purposes, respectively. The idle or "other" category represents fields growing something other than the specific listed categories (which included corn-soybeans, wheat-fallow, grass, mixed forage, and CRP). Note that the CRP category includes both General and Continuous Signup CRP, but the CRP acres shown after 2016 for fields rejected from the 2016 General Signup are enrolled through the Continuous Signup CRP. Finally, the "no majority crop" category represents fields where no specific category comprises more than 50 percent of the acreage in the years observed[92] (a complete set of values of land use proportions in the figure are provided in table A.1 of the appendix).

Overall, figure 2 shows that after being rejected from CRP, landowners put 16.6 percent of this acreage into corn and soybeans, 23.4 percent into wheat and fallow, 20.7 percent into grassland, 14.7 percent into mixed forage, and 8.9 percent into idle land and other land uses. From the pre-Signup to post-Signup periods, most landowners kept their non-CRP land in its pre-Signup uses after their offers were rejected. Nearly all acreage in grass or in mixed forage remained in those plantings; approximately three-quarters of acreage in wheat-fallow remained in wheat-fallow. Two-thirds of corn-soybean acreage remained in corn and soybeans. For those fields that were in CRP prior to applying in the 2016 Signup, the most common land use (after rejection) was grass, which is most often used for grazing. In general, post-Signup land uses of fields from returning CRP applicants were substantially different from those of new applicants. We elaborate on these differences in a later section.

Both returning and new applicants enrolled a substantial fraction of their acreage that was rejected from the 2016 General Signup in the CRP through Continuous Signup. The largest fraction of this enrollment was from the group with land exiting the CRP, but there was also a large proportion of corn-soybeans acreage that was enrolled in the Continuous Signup. It was somewhat surprising that so many landowners chose to make an offer through General Signup before going into CRP through Continuous Signup, given the potential for higher compensation from the latter. However, the cover practice required under Continuous Signup might be less desirable to some decision

[92] Specifically, the calculation is based on the fraction of acre-years within a field during the relevant period. For example, a field growing corn in 2017, growing wheat in 2018, and listed as idle in 2019 would be categorized as "no majority crop."

makers, or only a subset of their fields might be eligible for Continuous Signup.

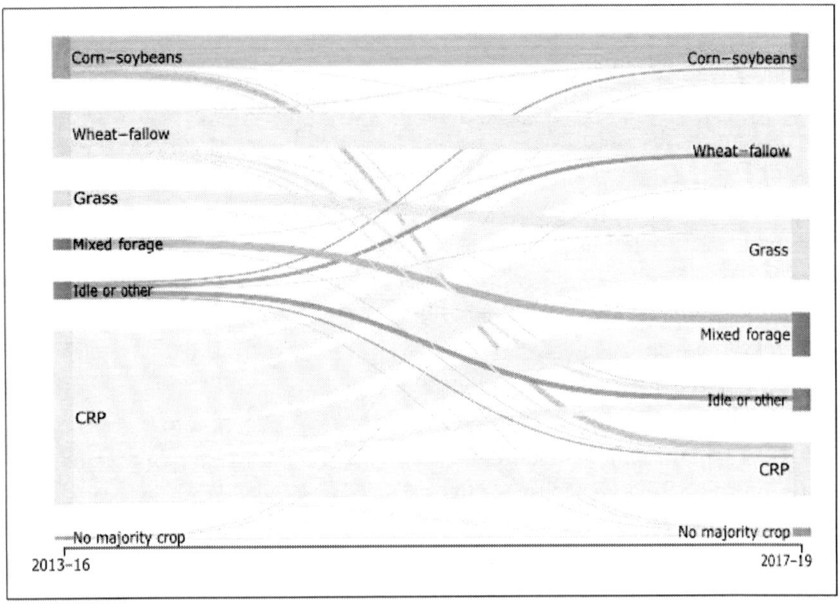

Source: USDA, Economic Research Service analysis of USDA, Farm Service Agency Crop Acreage Reporting Data and CRP offer data.

CRP=USDA's Conservation Reserve Program.

Note: Fields are assigned to a land-use category if the fields are in that land use for the majority of acre-years within each period. "Corn-soybeans" includes any corn or soybean plantings, including continuous corn, continuous soybean, and corn-soy rotations. "Wheat-fallow" includes any wheat or fallow, including continuous wheat, fallow, and wheat-fallow rotations. "Grass" and "mixed forage" include majority grass and mixed forage plantings, respectively. "CRP" includes fields with the majority of their land in CRP. "Idle or other" includes all other crop codes. "No majority crop" represents all acreage where none of these categories represents the majority of acre-years within the period. Data are shown for fields with rejected offers in the 2016 General Signup and no subsequent General Signup contract during the sample period. The thickness of bars reflects the proportion of acres planted to the associated crop, as reported to USDA, Farm Service Agency.

Figure 2. Land-use transitions of acres for fields rejected and not enrolled in a General Signup contract, before and after the 2016 General Signup.

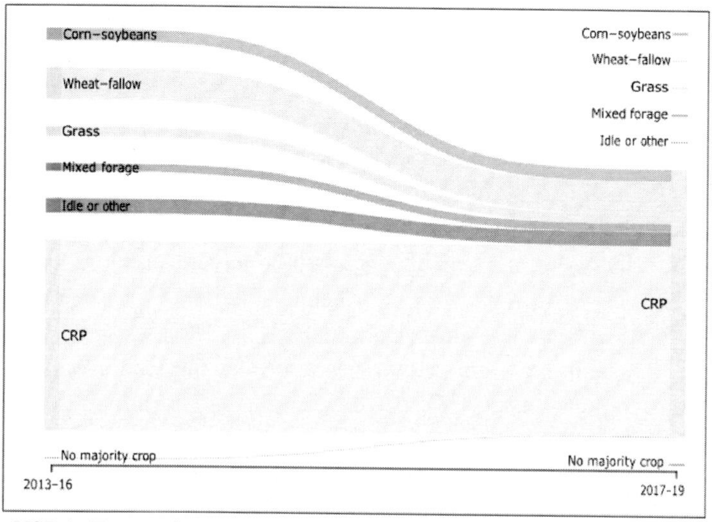

Source: USDA, Economic Research Service analysis of USDA, Farm Service Agency Crop Acreage Reporting data and CRP offer data.

CRP=USDA's Conservation Reserve Program.

Note: Fields are assigned to a land-use category if the fields are in that land use for the majority of acre-years within each period. "Corn-soybeans" includes any corn or soybeans, including continuous corn, continuous soybean, and corn-soy rotations. "Wheat- fallow" includes any wheat or fallow, including continuous wheat, fallow, and wheat-fallow rotations. "Grass" and "mixed forage" include majority grass and mixed forage plantings, respectively. "CRP" includes fields with the majority of their land in CRP. "Idle or other" includes all other crop codes. "No majority crop" represents all acreage where none of these categories represents the majority of acre-years within the period. Data are shown for fields with accepted offers in the 2016 General Signup. The thickness of bars reflects the proportion of acres planted to the associated crop, as reported to USDA, Farm Service Agency. Non-CRP land uses are rare for accepted and enrolled acres in 2017–19 but are possible if a field drops out of CRP or if the CRP contract exists on only a minority of the field, for example.

Figure 3. Land-use transitions of acres for fields accepted and enrolled in a General Signup contract, before and after the 2016 General Signup.

Figure 3 illustrates the land-use transitions for fields that were accepted in 2016 and enrolled in a General Signup contract. Unsurprisingly, nearly all of these fields were in a CRP land cover from 2017 to 2019. Note that the CRP category includes both General and Continuous Signup CRP, but the CRP acres shown after 2016 for fields accepted and enrolled into the 2016 General Signup were enrolled through General Signup CRP. The most common non-

CRP land use prior to the 2016 Signup among accepted acreage was the wheat/ fallow category. However, overall, the majority of accepted acreage was in a CRP land cover prior to the Signup. It is also true that returning CRP applicants were more likely to be accepted than new applicants.

There are two explanations for this difference. First, offers with higher EBI scores typically have more EBI points for immutable characteristics, such as erosion potential, location within priority zones, and estimated rental rates. As a result, offers with higher EBI scores would have been more likely to be accepted into CRP in previous years. Second, higher EBI scores were also associated with offers with more chosen EBI points, especially for cover practices.[93] This finding would likewise imply that the landowners offering these fields would likely have been more successful in previous efforts to enroll in CRP.

Figure 4 summarizes the proportions of acreage in different land uses across EBI levels in the 4 years prior to the Signup (top panel) and for the 3 years following the Signup (bottom panel) for all offers. In the figure, corn, soybeans, and wheat are separate categories, fallow is combined with idle, and a category for timber is included (previously included in "other"). Proportions for each category reflect the overall share of acreage in each category during the specified period for all acres with a given EBI score.[94] Offers are listed by which acreage is accepted first (that is, has the highest EBI scores), so that the figure depicts how land-use impacts vary as more acreage is hypothetically accepted into the program. Overall, in the 2016 Signup, landowners offered nearly 1.9 million acres in total, about 22 percent of which were accepted. After the process of identifying land uses for each field, we were left with about 1.5 million acres of offered land used throughout the analysis in this report.

The first dashed line in both panels of the figure represents the cumulative acreage where the 2016 EBI threshold of 292 falls, above which most offers were accepted. Dashed lines further to the right in the figure indicate where thresholds would have been in 2016 if the proportions of acreage accepted in other recent General Signups were accepted in 2016. Comparing the dashed

---

93 For a more thorough discussion of cover practices in the CRP General Signup, the incentives for choosing them, and the choices that landowners and producers make when offering to the General Signup, see Pratt and Wallander (2022).

94 For these categories, we are aiming for a more disaggregated approach as opposed to one that captures rotations. However, we wanted to minimize the number of categories with small proportions. Since idle and fallow are functionally similar and the amount of land categorized as "idle" is very small, we combined idle with fallow in this case.

lines in the figure, these other recent Signups accepted a much higher proportion of offered acreage than the 2016 Signup. As we argue in appendix B, the distribution of EBI scores and field characteristics of offers has remained similar across Signups. Thus, there is good reason to expect that land-use impacts across Signups would be similar for offers at similar points in the EBI distribution and that most rejected acres in 2016 were roughly representative of accepted acres in these other Signups.

The top panel of figure 4 shows land-use decisions in the immediate years prior to the 2016 Signup for all rejected and accepted offers. Interestingly, the pre-Signup land uses were somewhat stable across the full range of EBI scores. However, offers with lower EBI scores (where EBI scores were smaller as more acreage was accepted) tended to have a higher proportion of acres in corn or soybeans prior to making a CRP offer. Accepted offers were more likely to be in wheat or other crops prior to the Signup. For other land uses, the differences were small between the groups. For post-Signup land uses (bottom panel), the shares of acreage in most land uses were also stable across EBI scores of rejected offers. The fractions of acreage in corn and soybeans increased as EBI scores decreased, whereas the fractions in wheat and fallow/idle decreased. The proportions of land in grass, mixed forage, and other crops were mostly stable with EBI levels. Timber represented a small proportion of crops throughout the range of EBI. Finally, less than 20 percent of land went into CRP from 2017 through 2019 via Continuous Signup, with limited variation across EBI scores below the threshold (to the right of the first dashed line).

Figure 4 shows that land-use impacts varied across Environmental Benefits Index scores, which is largely due to the way in which the EBI formula is constructed. A total EBI score includes separate scores for six components, including air quality, enduring benefits, erosion, water quality, wildlife, and cost. These components include measures of the inherent properties of land being offered, as well as additional ways in which offers are made competitive (for instance, through rental rate discounts and practice choice). The cost component is a combination of scores from the maximum rental rate assigned to an offered parcel along with the discount offered by the landowner. In general, higher rental rates are penalized.

The EBI formula does not account for the specific land-use choices that are avoided when land is retired in CRP, though these choices are important for the environmental impacts of enrolled CRP land. For example, going from more intensive land uses like cropland to a CRP land cover will provide more environmental benefits than will going from grassland to CRP land cover for most types of environmental outcomes.

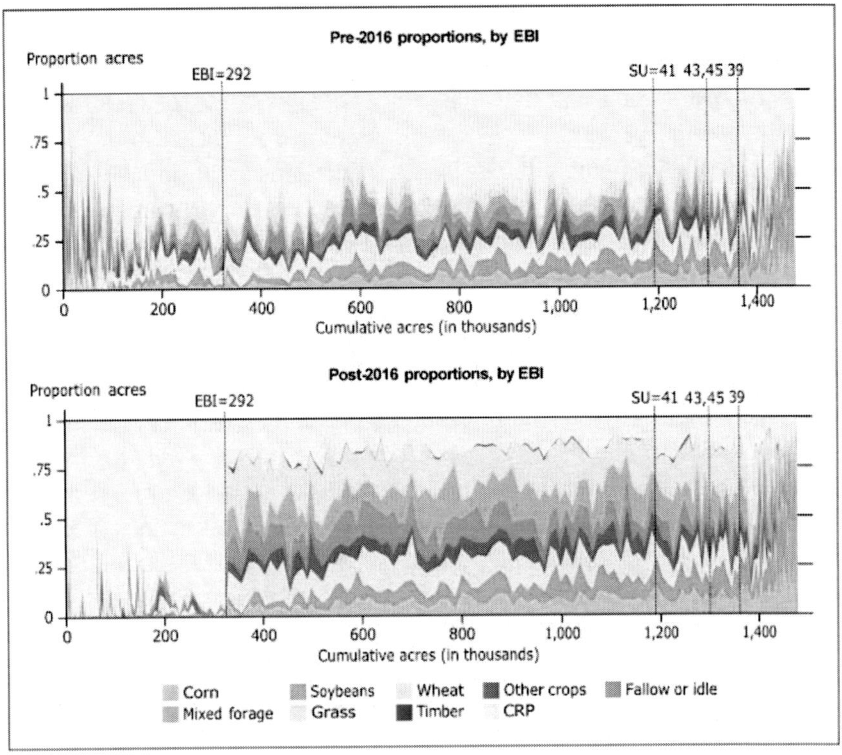

Source: USDA, Economic Research Service analysis of USDA, Farm Service Agency Crop Acreage Reporting data and CRP offer data.

CRP=USDA's Conservation Reserve Program. EBI=Environmental Benefits Index. SU = Signup.

Note: The top panel of the figure shows proportions of acres in each pre-Signup (2013–16) land use (delineated by color). The bottom panel of the figure shows proportions of acres in each post-Signup (2017–19) land use (delineated by color). Bar segment heights reflect the proportion of offered acreage in each land use. The 2016 EBI threshold, indicated with the left-most dashed line in each panel of the figure, was 292. Amounts of cumulative acres from 2016 that are proportional to the amounts accepted in other Signups are also indicated with dashed lines, with the Signup numbers indicated.

Figure 4. Pre-Signup and post-Signup land use of offers.

However, although the EBI formula does not incorporate land use changes induced by CRP explicitly, the penalty for higher rental rates induces a relationship. Higher rental rates correspond with a higher likelihood of crops being grown; thus, the EBI indirectly penalizes land in more intensive land uses.

Land-Use Impacts Per Hundred Dollars Spent

In considering the land-use impacts of the CRP and correlations between the EBI and land-use impacts, cost is another important component. Retiring land with a higher proportion in crops may bring higher environmental benefits per acre, but crop acreage comes at a higher cost to CRP on average. Land in commodity crops generally rents for higher rates compared to land in grazing or other less intensive uses. Further, USDA, FSA constrains offers to rental rates at or below a maximum rate based on soil and county characteristics.[95] Given these factors, a tradeoff exists between the higher environmental benefits afforded by retiring more intensive land uses and the increased costs of retiring that land. This section accounts for these competing factors, examining the acres of each land use that would have been retired per $100 spent on rejected offers in the 2016 General Signup (tables providing precise numbers for the graphs presented in this section are available in appendix A).

Figure 5 illustrates the average impact of accepting all rejected offers calculated per additional $100 spent on CRP rental payments annually, excluding those offers that enrolled in the CRP through Continuous Signup. For these rejected offers, a total of 1.39 acres would be retired per $100 spent on rental payments annually.

Among these acres, the largest portion of acreage of land retired would go into grass (0.33 acres per $100), followed by mixed forage (0.24 acres). Further, on average, an additional $100 in rental payments would also have retired 0.23 acres of land in wheat, 0.15 acres in soybeans, 0.13 acres of land

---

[95] The maximum rental rate is intended to be an estimation of the opportunity cost for the field. At the time of the 2016 General Signup, this maximum rental rate was determined by the multiplicative combination of the following: the county average dryland cash rental rate, as determined by USDA, National Agricultural Statistics Service surveys; a soil productivity factor that scales based on the ratio of the field's National Commodity Crop Productivity Index and the county average National Commodity Crop Productivity Index; and an inflationary factor of 10 percent. Although not in effect for the studied Signup, the 2018 Farm Bill restricted maximum rental rate, which is currently executed with a multiplicative factor of 0.85 times the county rental rate.

in corn, 0.12 acres in other crops, and 0.18 acres in fallow or idle use. Only about 0.01 acres per $100 from rejected offers would have gone into retiring timberland. Note that these impacts are averages split across many types of parcels and any particular parcel retired will in most cases be concentrated in only one land use.

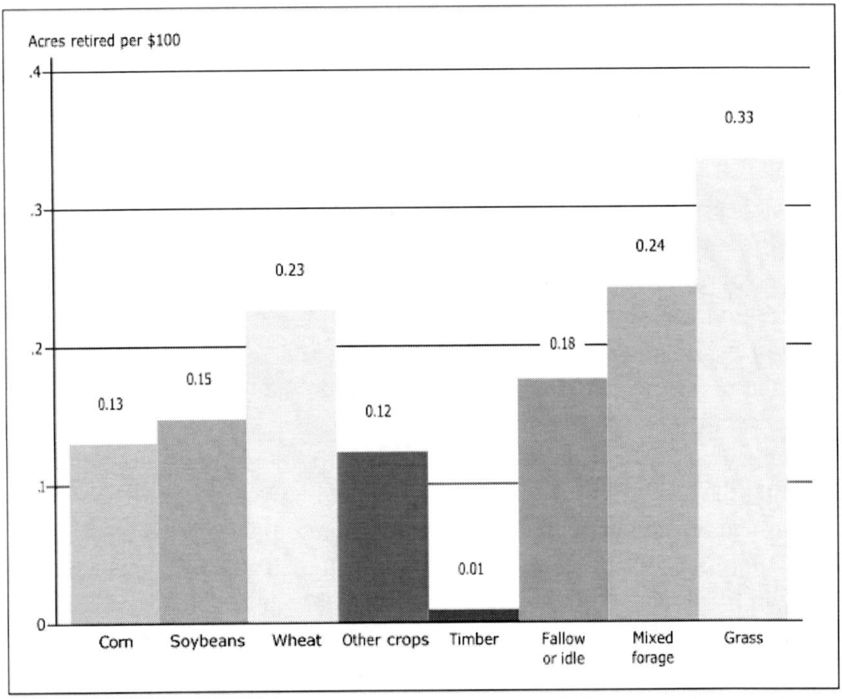

Source: USDA, Economic Research Service analysis of USDA, Farm Service Agency Crop Acreage Reporting data and CRP offer data.

CRP=USDA's Conservation Reserve Program.

Note: This figure shows acreage in each post-Signup (2017–19) land use (delineated by color) that could be retired per $100 in rental rates if offers rejected during the 2016 Signup were accepted. Specific values for the acres in each post-Signup land use are reported in the figure. Acres potentially retired per $100 are calculated by dividing the proportions of acres in each post-Signup land use by total offered annual rental payments of rejected offers. Fields going into Continuous Signup CRP are excluded. Bar heights reflect the average of offered acreage in each land use per $100 in potential rental payments.

Figure 5. Acres potentially retired per $100 spent on rejected offers (excluding Continuous Signup CRP).

The cost of retiring acreage across different land uses can vary with how much acreage is enrolled in the program. We show this relationship using a similar graphic to figure 4, but we summarize the average acreage that went into each land-use category for every $100 that would have been paid in annual rental rates if the offers were accepted. Figure 6 shows how the potential cost of enrolling rejected land into the CRP varies with EBI scores, and in effect, the total acreage accepted into the program. The horizontal axis is still cumulative acres but with some limitations. First, these are cumulative acres for only rejected offers. Second, this does not include land that goes into the Continuous Signup. Thus, here we only assessed how cost of retirements varies for land on which we could directly observe non-CRP land-use choices. Together, this accounted for about 950 thousand acres of the 1.45 million acres of offered land included in the analysis. Figure 6 also provides the average Environmental EBI score at each EBI level. The Environmental EBI is defined as the total EBI score minus the score assigned to the cost component, following USDA, FSA's approach for measuring potential environmental benefits of CRP enrollments. The figure shows that Environmental EBI scores decrease as more acreage is hypothetically accepted into the program. Thus, even if acres retired per $100 do not change, the estimated potential environmental benefits of offers decrease as more acreage was enrolled into the program.

In figure 6, acres retired per hundred dollars in annual rental payments are broadly decreasing as more acreage is accepted into CRP, that is, increasing in EBI scores. This results because the CRP rewards more EBI points to offers with lower rental rates. However, the relationship was approximately flat across more than half of the distribution, reflecting the concentration of offers in a small range of EBI scores. Acreage in corn and soybeans increased substantially as EBI values decreased. On the other hand, acreage in grass, mixed forage and fallow/idle uses decreased and constituted the bulk of differences across EBI scores.

Figure 6 provides evidence that the crop acreage potentially retired in CRP per hundred dollars spent is fairly stable as more acreage is enrolled into the program. This finding occurs because the EBI penalty for higher rental rates relates to acres per hundred dollars in two directions. First, the penalty directly rewards parcels that can be enrolled more cheaply, meaning offers with more acreage retired per $100 have higher EBI scores. Second, as figure 4 shows, there is a negative correlation between proportion of crop acreage and the EBI.

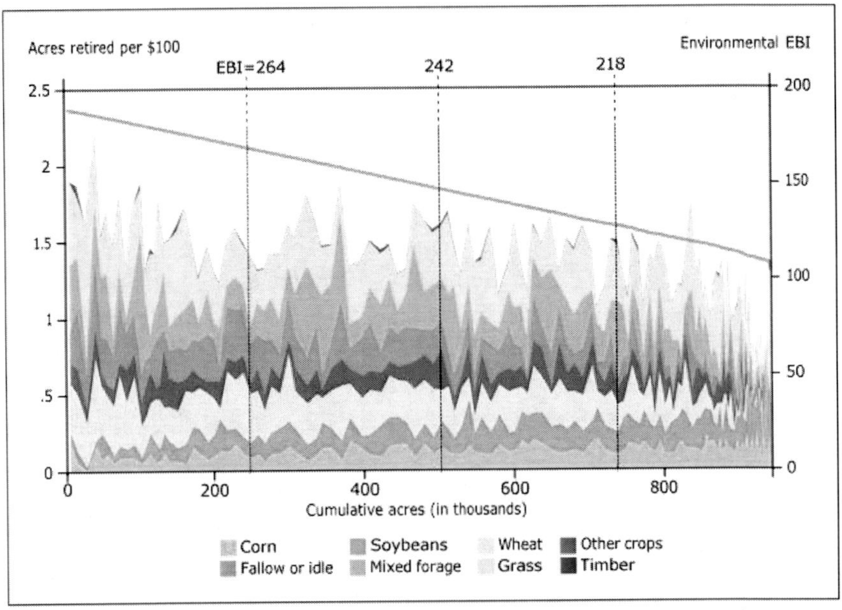

Source: USDA, Economic Research Service analysis of USDA, Farm Service Agency Crop Acreage Reporting data and CRP offer data.

CRP=USDA's Conservation Reserve Program. EBI=Environmental Benefits Index.

Note: This figure shows acreage in each post-Signup (2017–19) land use (delineated by color) that could be retired per $100 in rental rates if offers rejected during the 2016 Signup were accepted by EBI. Offers are ordered in decreasing EBI levels from left to right, and the horizontal axis is scaled to reflect the cumulative acres enrolled by accepting all offers between the true 2016 EBI threshold (292) and the EBI value in question. The amount of acres potentially retired per $100 is calculated by dividing the total acres in each post-Signup land use by total rental payments across all rejected offers for each EBI value. Fields going into the Continuous Signup CRP are excluded. Area heights reflect the offered acreage that would be retired per $100 in potential rental payments for each EBI value of rejected offers. The total of acres retired per $100 is shown by the uppermost edge of each area graph. The overlying lines depicted show how the Environmental EBI, defined as the noncost portion of EBI, varies by EBI levels (or cumulative acres).

Figure 6. Acreage of land uses potentially retired per $100 spent by EBI (excluding Continuous CRP).

Variation in the Land Uses of New and Returning Applicants

This section examines how land-use impacts of CRP differ based on the prior CRP experience of land- owners making CRP offers. As figures 2 and 3 show, a large proportion of offers in the 2016 General Signup included expiring CRP land. There are theoretical and empirical reasons to believe that returning and new applicants will be in differing land uses after rejection from the General Signup. Land enrolled in CRP is planted to a perennial cover that may include a diverse array of species and plant types. As discussed in Pratt and Wallander (2022), the typical cost of establishing such cover for popular practices can be in the range of $50 to $100 per acre. Accordingly, landowners may prefer to utilize this cover for grazing or other purposes rather than immediately destroying the cover to plant a commodity crop. Compared with the overall costs of converting the land to commodity crop production, using the land for grazing may be more profitable. Landowners are more likely to persist in their existing use than to transition unless entering their fields into CRP as a new enrollment.

Figure 7 shows the post-2016 land uses of rejected offers conditioned on prior land use. The left column in the figure shows the land-use decisions made by landowners with rejected offers without prior CRP enrollment, that is, new applicants with rejected offers. The middle column in the figure shows land uses of landowners with rejected offers that had land in CRP that expired before 2016, and the right column presents land uses of landowners with rejected offers with land exiting CRP in 2016. New applicants were substantially less likely to enter CRP through Continuous Signup (after being rejected from General Signup) than either group of returning applicants. Landowners with CRP contracts expiring in 2016 were much more likely to enter Continuous Signup after being rejected than either of the other groups. We could also see more clearly here the substantial difference between new and returning applicants in the amount of their land they put into crops after being rejected, with new applicants much more likely to go into corn, soybeans, and wheat.

Less than 20 percent of acres from returning applicants with CRP expirations in 2016 were in crops of any kind. Further, both groups of landowners with prior CRP enrollment had similar proportions of acreage in grass or mixed forage when Continuous Signup land is excluded (61 percent in both cases). These levels differed dramatically from the land-use decisions of landowners with no prior CRP enrollment. The most plausible explanation is that returning applicants did not quickly remove the CRP plantings either between expiration and the 2016 General Signup or after being rejected from that Signup. CRP plantings primarily comprise grasses and plants that may provide value as forage, which could explain why these plantings would persist beyond contract expiration for the majority of fields. Providing further evidence that fields kept their land in conservation cover, a large minority of the grassland and mixed forage land managed by returning applicants were categorized in CARD as "left standing" during 2017 through 2019—more than 30 percent of the acreage in grass and more than 15 percent of the acreage in mixed forage. Crops that were left standing were indicated as having no commercial purpose. The majority of grassland was used for grazing, while the percentage for mixed forage was relatively evenly split, with 40 percent each for grazing and forage (table A.6). Figure E.3 in the appendix shows how land-use impacts vary for new and returning applicants across space.

We next explore how the land uses that could be retired per $100 spent on enrolling additional rejected acreage differ by prior CRP participation. Figure 8 illustrates the post-Signup land uses of landowners with rejected offers across EBI scores by prior CRP status. The results in the figure largely reflect the differences in the proportions of land use in figure 7. Comparing the top panel of figure 8 with the bottom panel, retiring additional acres from returning applicants would lead to significantly different types of land being retired.

Additionally, enrolling these offers would retire substantially more acres in grassland per $100, as well as more acres in mixed forage. On the other hand, enrolling new applicants (in the top panel) would retire substantially more acres in crops per $100. For fields both with and without prior CRP enrollment, total acreage in crops tended to increase in EBI scores, although acreage in corn and soybeans decreased.

Figure 8 also demonstrates that, while less total land was retired per $100 for new applicants, more acres of crops (corn, soybeans, wheat, and other crops) were retired per $100 for new applicants compared to returning applicants (for comparable overall EBI levels and Environmental EBI levels). Whether it is desirable to prioritize newly enrolled land over returning applicants may depend on program priorities. For certain benefits (such as wildlife habitat or carbon sequestration), there may be reason to prioritize long-term enrollment, even if the land being displaced is less likely to be used intensively. For example, crop acreage is more likely to be tilled, which can release large quantities of carbon stored for many years back into the atmosphere (Li et al., 2022; Sullins et al., 2021). On the other hand, different environmental benefits may require that the most environmentally sensitive acres be replaced with a conservation cover annually. In that case, new applicants would likely bring greater benefits per dollar spent.

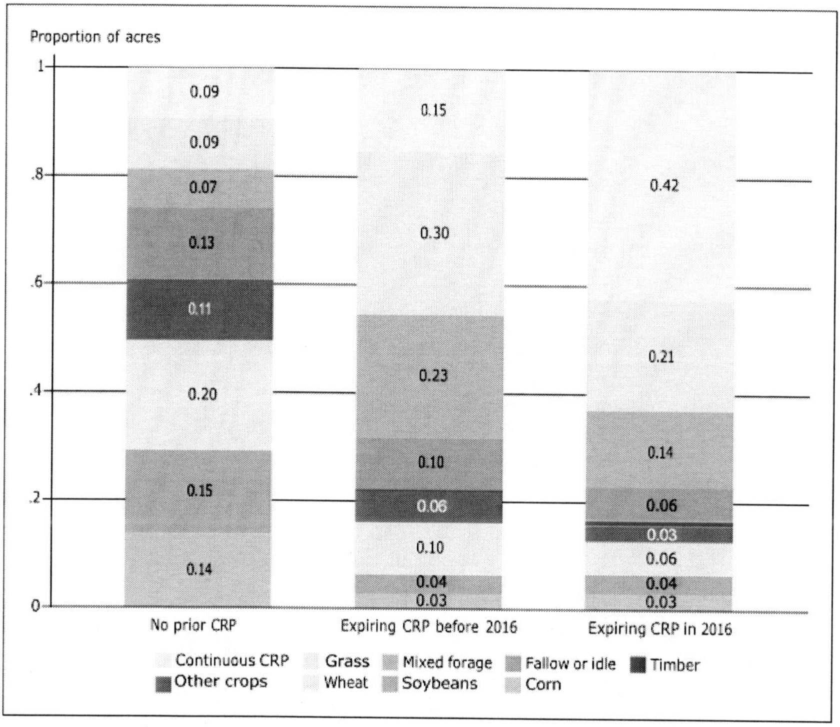

Source: USDA, Economic Research Service analysis of Farm Service Agency Crop Acreage Reporting Data and CRP offer and contract data.

CRP=USDA's Conservation Reserve Program.

Note: This figure shows proportions of acres in each post-Signup (2017–19) land use (delineated by color) for rejected offers, split by whether the offered land was in a CRP contract that expired during 2016, in a contract that expired between 2013 and 2015, or not in a CRP contract between 2013 and 2016. Specific values for the proportion of acres in each post-Signup land use are reported in the figure. However, values under 0.03 are not shown due to lack of space. Proportions across categories add to 1, but reported values may not due to omitted values or rounding. Bar segment heights reflect the proportion of offered acreage in each land use.

Figure 7. Post-Signup land use of offers rejected in the General Signup, by prior CRP status.

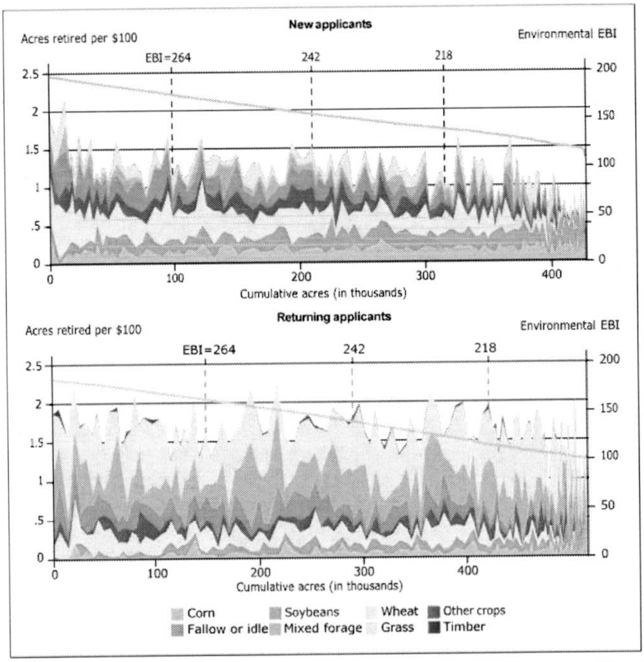

Source: USDA, Economic Research Service analysis of USDA, Farm Service Agency Crop Acreage Reporting data and CRP offer data.

CRP=USDA's Conservation Reserve Program.

Notes: This figure shows acreage in each post-Signup (2017–19) land use (delineated by color) that could be retired per $100 in rental rates if offers rejected during the 2016 Signup were accepted by Environmental Benefits Index (EBI) and prior CRP status. The top panel includes only fields with rejected offers and no CRP during the period of 2013 through 2016. The bottom panel includes only fields with rejected offers and some amount of CRP during the period of 2013 through 2016. Offers are ordered in decreasing EBI levels from left to right, and the horizontal axis is scaled to reflect the cumulative acres enrolled by accepting all offers between the true 2016 EBI threshold (292) and the EBI value in question. Acres potentially retired per $100 are calculated by dividing the total acres in each post-Signup land use by total rental payments across all rejected offers for each EBI value and offer group. Fields going into Continuous Signup CRP are excluded. Area heights reflect the offered acreage that would be retired per $100 in potential rental payments for each EBI value of rejected offers and offer group. The total of acres retired per $100 is depicted by the upper- most edge of each area graph. The overlying lines depicted show how the Environmental EBI, defined as the noncost portion of EBI, varies by EBI levels (or cumulative acres).

Figure 8. Acres potentially retired per $100 in each land use by prior CRP status, for fields rejected from the 2016 CRP General Signup.

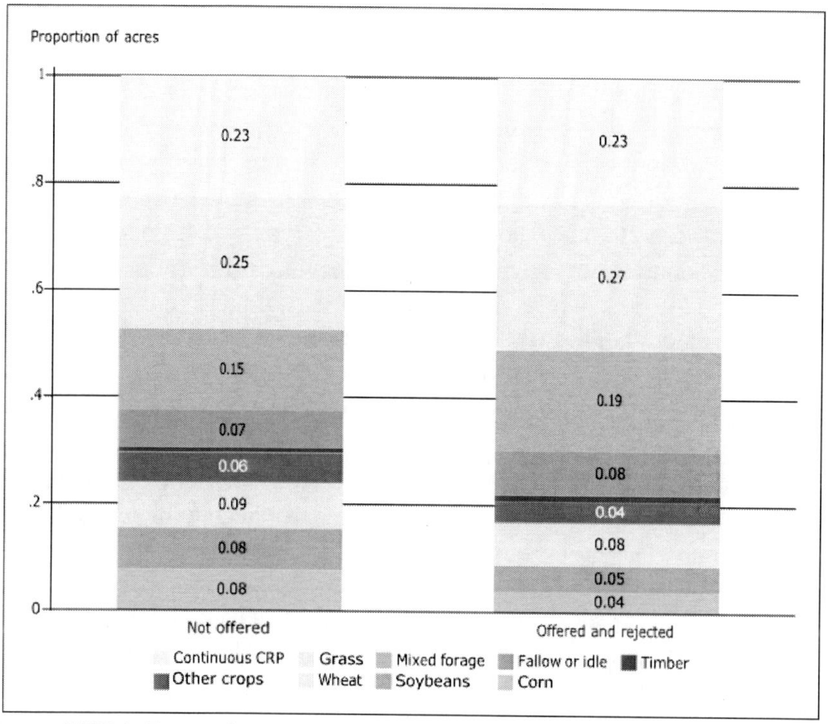

Source: USDA, Economic Research Service analysis of USDA, Farm Service Agency Crop Acreage Reporting data, CRP offer data, and contract data.

CRP=USDA's Conservation Reserve Program.

Note: This figure shows post-Signup (2017–19) land use weighted by acreage of fields, including only those fields with an expiring CRP contract in 2016 with a "General Signup" practice. Specific values for the proportion of acres in each post-Signup land use are reported in the figure. However, values under 0.03 are not shown due to lack of space. Proportions across categories add to 1, but reported values may not due to omitted values or rounding. Expiring CRP is restricted to conservation practices (CPs) found in General Signup contracts. Fields may have been in a Continuous Signup contract with one of these practices, but Continuous Signup contracts with these practices would represent a small portion of the fields included. An individual field may contribute to more than one land-use category, with the appropriate fraction apportioned to each category. The "Continuous CRP" category refers to Continuous Signup enrollment.

Figure 9. Post-2016 land use of fields with General CRP practices expiring in 2016, by offer status and excluding 2016 General Signup enrollments.

Comparing Voluntary and Involuntary Exits

This section compares land-use decisions of landowners with expiring CRP land who attempted to reenroll in CRP and were rejected with those with expiring CRP who did not attempt to reenroll. Figure 9 provides a general overview of land-use outcomes between these two groups. The left bar of the figure shows post- Signup land-use decisions of landowners with fields in General CRP contracts set to expire on September 30, 2016, who did not make an offer in the 2016 General Signup. The right bar of the figure shows post-Signup outcomes of landowners with fields in General CRP contracts expiring in 2016 who made a rejected offer in the 2016 General Signup. First, it is apparent that those landowners who did not make an offer were just as likely as those who did to end up in CRP through Continuous Signup. However, overall preferences varied between these groups. The nonoffering group found either Continuous Signup or not participating in CRP at all to be preferable to General Signup. The group that made offers to reenroll preferred General Signup. The second main finding shown in figure 9 is that those not making an offer were more likely to end up planting crops and less likely to end up in grasses or mixed forage.

Figure C.1 in the appendix presents a similar comparison to that in figure 9 but removes those fields going into CRP through Continuous Signup. The appendix figure C.1 provides a comparison of true exits from the CRP. The left bar shows those landowners "voluntarily exiting" CRP—those with expiring CRP contracts in 2016 who did not apply for General Signup in 2016 and who also did not enroll through Continuous Signup from 2017 through 2019. The right bar shows landowners that were "involuntarily exiting" CRP—those with expiring CRP contracts in 2016 who had offers rejected from the 2016 General Signup and who were not enrolled in CRP in 2017 through 2019. Landowners who voluntarily exited were more likely to go into corn, soybeans, and other crops—while landowners who involuntarily exited CRP were more likely to go into grass or mixed forage. The differences in corn and soybeans, in particular, were statistically significant even when controlling for prior land use and local differences.

These results mirror those from Barnes et al. (2020), who also found that landowners voluntarily exiting CRP were more likely to go back into crops. These results could be due to underlying differences in land quality as well as other factors; for example, landowners wanting to grow crops on the land will be more likely to exit the CRP voluntarily. Appendix C also looks at post-2016 land use of fields with expiring CRP, including exits from Continuous

Signup, in figures C.2 and figure C.3. The primary difference is that substantially more acreage among the group not making a General Signup offer enrolled in CRP through Continuous Signup.

This finding may reflect landowners of these fields being disproportionately eligible for and interested in the Continuous Signup. Appendix D explores the comparison between rejected offers and nonoffered land in greater depth.

Geographic Differences in Land-Use Decisions of Rejected Offers

Another important way in which land-use impacts varied was geographical. Figure 10 illustrates the proportion of acres in each land use during 2017 through 2019 among fields rejected from the 2016 General Signup, by U.S. State. The proportions for non-CRP land use across States largely reflect the predominant land uses in those States. For example, Corn Belt States like Iowa, Illinois, and Minnesota have a high proportion of land in corn and soybeans after being rejected. On the other hand, drier States like Oklahoma and New Mexico have a lot of rejected acres in grassland. Figure 10 also emphasizes that there was a large amount of variation across States in the proportion of land that eventually enrolled in the CRP through Continuous Signup. Large portions of acreage entered the Continuous Signup in Mississippi, Louisiana, and Iowa, among other States. Values of proportions in figure 10 are provided in table A.2 of the appendix.

Midwestern States like Illinois, Iowa, and Missouri had high proportions of land in corn and soybeans after being rejected. Kansas and North Dakota had large proportions of land in wheat production after rejection. Several States in the South had a large proportion in timber, including Louisiana, South Carolina, Mississippi, North Carolina, Alabama, and Arkansas, even though timber represents a low proportion of overall acreage for rejected offers. Proportions from figure 10 (not including land in Continuous Signup) are provided in table A.3 and figure E.1 of the appendix.

Figure 11 provides a clearer picture of the proportions in each land use for parcels rejected from CRP that were not subsequently enrolled into CRP. The figure provides maps for all crops (panel A), which combines the four categories of crops in figure 10: mixed forage (panel B), grassland (panel C), and fallow or idle land (panel D). As the figure shows, the proportion of crop acreage in Midwestern States was very high for those landowners who did not enroll into CRP. Grassland was most common for rejected land in the Southern High Plains, and mixed forage was most common in the Northern High Plains.

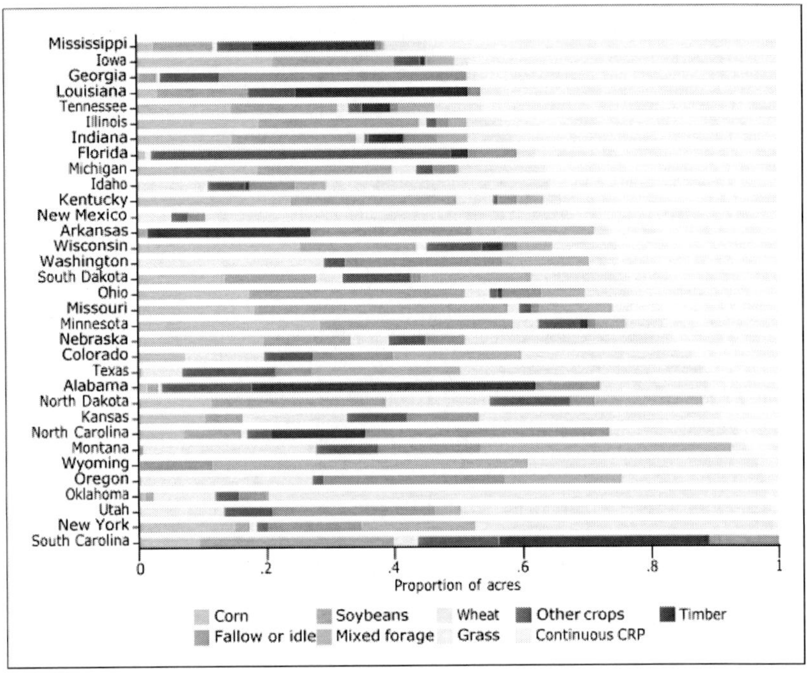

Source: USDA, Economic Research Service analysis of USDA, Farm Service Agency Crop Acreage Reporting data and CRP offer data.

CRP=USDA's Conservation Reserve Program.

Note: This figure shows the proportions of acres in each post-Signup (2017–19) land use (delineated by color) for rejected offers by U.S. State. Bar segment widths (in the horizontal dimension) reflect the proportion of offered acreage in each land use within a State. States are ordered by proportion going into Continuous Signup CRP, starting with the highest proportion. Not all States are included in the analysis. Some States are omitted due to either insufficient or no offers submitted in 2016, or lack of observations after linking offers to land use data.

Figure 10. Proportion of acres in land uses for rejected offers, after CRP Signup 49, by U.S. State.

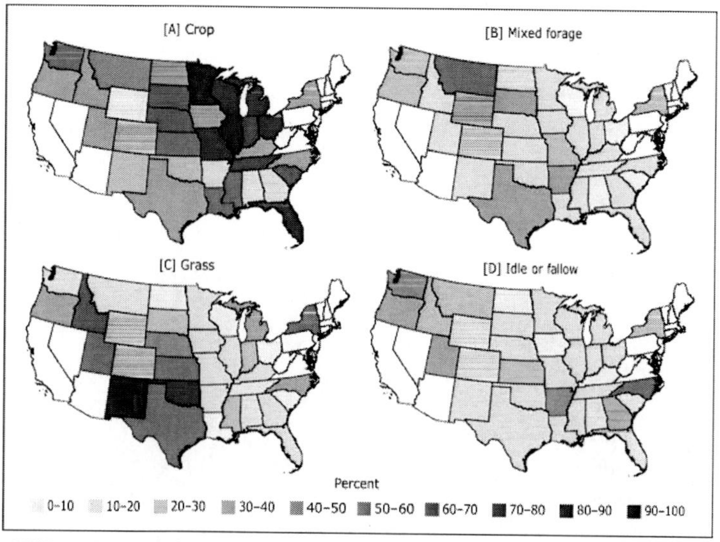

Source: USDA, Economic Research Service analysis of USDA, Farm Service Agency Crop Acreage Reporting data and CRP offer data.

CRP=USDA's Conservation Reserve Program.

Note: Each panel illustrates the percentage of acres offered to and rejected from the 2016 General Signup by State. Panel A shows the percentage of this acreage that subsequently was used for commodity cropping. Panel B shows the percentage used for mixed forage. Panel C shows the percentage in grass. Panel D provides the percentage of idle or fallow. Alaska and Hawaii are visually omitted from all panels, as there were no fields offered in Signup 49 from these States. States in white are omitted due to either insufficient or no offers submitted in 2016, or a lack of observations after linking offers to land use data.

Figure 11. Land-use percentages during 2017 through 2019 among fields rejected from the CRP 2016 General Signup.

Figure 12 shows the acres in each land use that would have been retired per $100 had rejected offers instead been accepted, by State. The figure shows that retiring more land in Wyoming, Texas, Montana, Colorado, New Mexico, and Utah all would have provided an average of nearly three or greater than three acres retired per $100 spent on enrolling rejected offers. However, in each of these States, the majority of the acres were planted in grass, mixed forage, or fallow/idle after rejection from the 2016 General Signup. States that had the lowest amounts of total acres per $100 spent, tended to have the highest amounts in corn and soybeans (such as Minnesota, Iowa, Indiana, and Illinois). However, although States in the Corn Belt tended to have high amounts of crop acreage for grain, these States retired less crop acreage per

$100 spent. Other States that had high amounts of wheat and other crops had the most potential acres retired in crops overall per $100 (like Kansas, Montana, and North Dakota). As we found with the proportions of land use, southern States like South Carolina and Louisiana had high acreages that went into timberland after being rejected (values of the proportions in figure 12 are provided in table A.4 of the appendix).

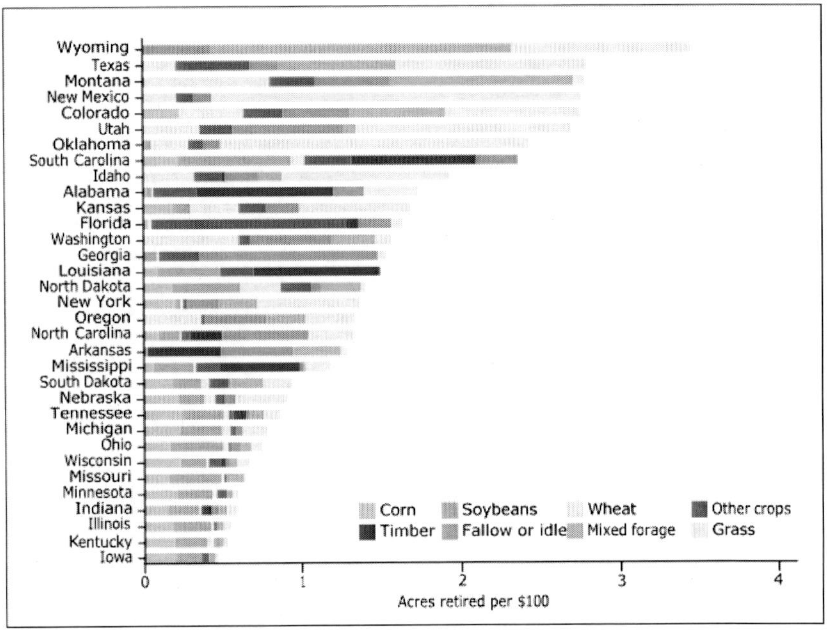

Source: USDA, Economic Research Service analysis of USDA, Farm Service Agency Crop Acreage Reporting data and CRP offer data.

CRP=USDA's Conservation Reserve Program.

Notes: This figure shows acreage in each post-Signup (2017–19) land use (delineated by color) that could be retired per $100 in rental rates if offers rejected during the 2016 Signup were accepted by U.S. State. Acres potentially retired per $100 are calculated by dividing the proportions of acres in each post-Signup land use by total offered annual rental payments of rejected offers for each State. Fields going into Continuous Signup CRP are excluded. Not all States are included in the analysis. Some States are omitted due to either insufficient or no offers submitted in 2016, or lack of observations after linking offers to land use data. Bar segment widths reflect the offered acreage in each land use per $100 in potential rental payments for each State.

Figure 12. Acres potentially retired per $100 spent on rejected offers, by U.S. State (excluding Continuing CRP Signup).

Figure 13 shows how acres per $100 varied by land use for cumulative acres by region. The authors chose six different regions from among ERS resource regions (appendix figure E.4 depicts all nine resource regions). The figure shows how much acres would be retired per $100 spent by region, at different levels of the EBI. For figure 13, we used the simplified land-use categories described for the maps of figure 11. The figure shows that the Heartland (which encompasses States like Iowa and Illinois) was mostly acreage in crops retired, but few acres were retired per $100 across the Heartland EBI spectrum. On the other hand, enrolling land in the Northern Great Plains and Prairie Gateway tended to have a similar amount of acreage in crops enrolled per $100 but with more of other types. Patterns in the Fruitful Rim, Basin and Range, and other regions tended to be more irregular.

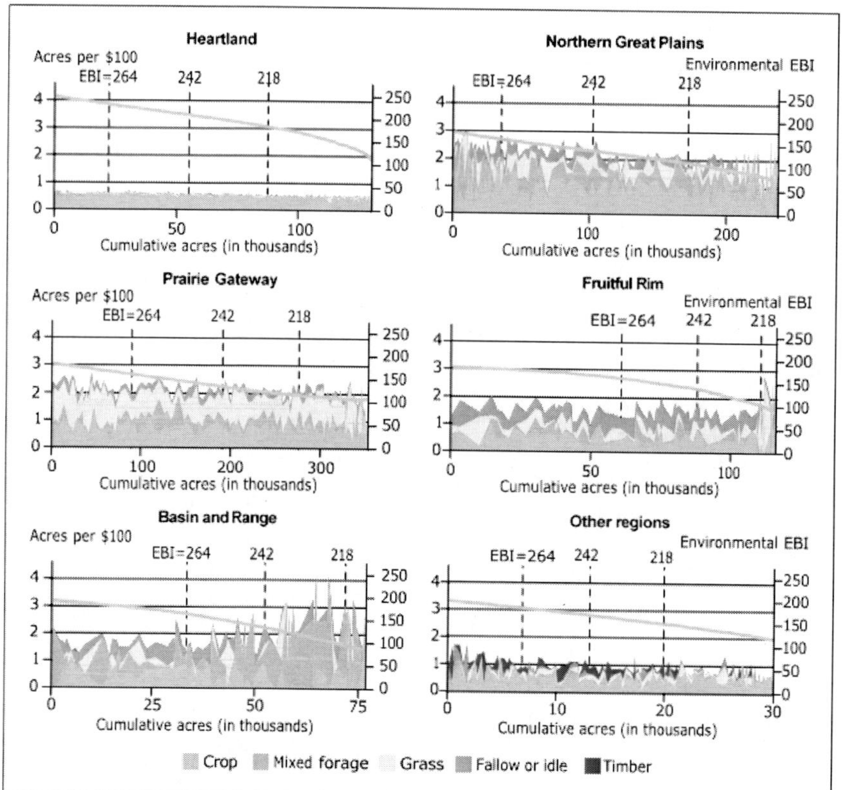

Source: USDA, Economic Research Service analysis of USDA, Farm Service Agency Crop Acreage Reporting data and CRP offer data.
CRP=USDA's Conservation Reserve Program.

Note: The figure shows acreage in each post-Signup (2017–19) land use (delineated by color) that could be retired per $100 in rental rates if offers rejected during the 2016 Signup were accepted, by Environmental Benefits Index (EBI) and USDA, ERS region. The acres retired per $100 is indicated by "Acres per $100." Offers are ordered in decreasing EBI levels from left to right, and the horizontal axis is scaled to reflect the cumulative acres enrolled by accepting all offers between the true 2016 EBI threshold (292) and the EBI value in question. Acres potentially retired per $100 are calculated by dividing the total acres in each post-Signup land use by total rental payments across all rejected offers for each EBI value and USDA, ERS region. Fields going into Continuous Signup CRP are excluded. Area heights reflect the offered acreage that would be retired per $100 in potential rental payments, for each EBI value of rejected offers and USDA, ERS region. Total acres retired per $100 are given by the uppermost edge of each area graph. The overlying lines depicted show how the Environmental EBI, defined as the noncost portion of EBI, varies by EBI levels (or cumulative acres).

Figure 13. Acres potentially retired per $100 in each land use by EBI and region, fields rejected from 2016 CRP General Signup.

These regional differences bring tradeoffs of their own. Although at first glance the Heartland appears to be the least effective at enrolling land likely to bring environmental benefits, Environmental EBI scores tended to be higher for given total EBI levels. Thus, enrolled land in this region was expensive but brought high environmental benefits due to high crop acreage intensity and high environmental scores. On the other hand, the Northern Great Plains and Prairie Gateway tended to have more overall land enrolled per $100. Finally, it was interesting to see the differences in the proportion of offered acreage in each region that fell below certain EBI levels. For example, a large proportion of land in the Fruitful Rim had high EBI scores. The dashed line indicating an EBI level of 264 was far to the right in the graph. This finding is also reflected in the high amount of acres enrolled in this region in Signup 49, particularly in eastern Washington. This information indicates how the program currently prioritizes offers across regions.

Post-Signup Land-Use Decisions by Offered Practice

In most of this report, the authors assess the land-use impacts of CRP as the non-CRP land use that is being displaced by CRP when landowners enroll their land. However, CRP impacts also depend on the specific CRP cover chosen when land is enrolled. Thus, to assess CRP impacts, we took advantage of the information available about both the proposed CRP land cover and the actual land-use outcome for rejected offers. Figure 14 provides an overview of land uses of all rejected offers in the years following the 2016 Signup. In the figure, shares of acreages by land-use categories are shown across broad CRP practice categories. Cover practices in General Signup CRP can be categorized broadly as grass, habitat for rare and declining species and pollinator habitat (collectively called "rare"), wildlife habitat (called "wild"), and trees.[96] The size of each rectangle is weighted by the proportion of total acreage represented.

As is visible along the horizontal axis, the majority of offered acreage was in grass practices, whereas rare and declining habitat and wildlife practices were less common. Tree practices comprised the least amount of acreage. Among offered grass practices, nearly 40 percent went into some type of crop after rejection. Figure 14 suggests that acreage offered to rare habitat and wildlife practices (which each contribute higher amounts to EBI scores) was associated with a higher proportion in intensive crops. This finding results in part because offers from parcels with higher rental rates must offer more competitive practices to make up for lower baseline EBI scores. Perhaps unsurprisingly, a large proportion of acreage that offered tree practices went into timber. This acreage primarily comprised land that was previously in CRP. Further, for the nontree practices, nearly 20 percent of acres found another way into CRP, likely through Continuous Signup.[97] The proportion entering CRP through Continuous Signup was nearly 40 percent for land that offered tree practices (see additional figures that split figure 14 by prior CRP status in appendix G).

---

[96] Tree practices constitute less than 700,000 acres nationally and geographically concentrated outside our sample. Pollinator habitat constitutes approximately 540,000 acres nationally. Both tree practices and pollinator habitat are excluded from the figure, given their limited acreage.

[97] The vast majority of cover practice codes among these fields are only available through Continuous Signup, and those cover practice codes that are available through General Signup are relatively rare in Continuous Signup. Grasslands Signup uses a separate set of cover practice codes. Because these cover practice codes are reported in the planting data that we use, we can identify CRP acreage on these fields as either definitely Continuous, probably General, or Grasslands Signup. Notably, almost all of this acreage is in cover practice codes that are only available through Continuous Signup, suggesting either preexisting partial-field enrollments or new enrollments in Continuous Signup. There may be preexisting General Signup partial-field enrollments on fields with rejected offers, but the evidence would suggest this is rare.

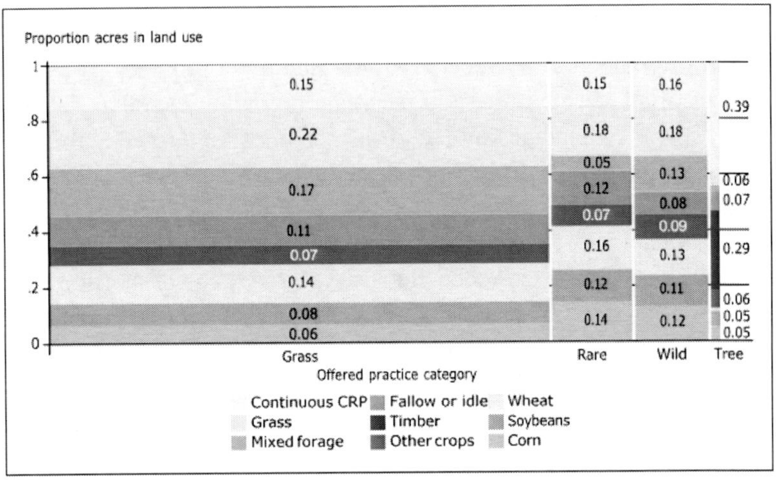

Source: USDA, Economic Research Service analysis of USDA, Farm Service Agency Crop Acreage Reporting data and CRP offer data.

CRP=USDA's Conservation Reserve Program.

Note: This figure shows the proportions of acres in each post-Signup (2017–19) land use (delineated by color) for rejected offers by offered practice category (horizontal axis). Specific values for the proportion of acres in each post-Signup land use are reported in the figure. However, values under 0.03 are not shown due to lack of space. Proportions across categories add to 1, but reported values may not due to omitted values or rounding. Bar widths for each practice category (in the horizontal dimension) reflect the proportion of total offered acreage in each offered category. For example, the majority of rejected acreage had offered to implement a grass practice. Bar segment heights (in the vertical dimension) reflect the proportion of acres, within offered practice category, in a given land use or crop. The grass practice category includes "Introduced grasses" and "Native grasses." The rare practice category includes "Rare and declining habitat" and "Pollinator habitat." The wild practice category includes "Wildlife habitat." The tree practice category includes "Tree planting" and "Hardwood tree planting."

Figure 14. Proportion of acres in land uses by practice category of 2016 General Signup rejected offers.

Conclusion

The CRP General Signup aims to be cost-effective at enrolling environmentally beneficial land. The program prioritizes lower cost land and land estimated to have the highest potential environmental impacts but does not account for the actual land-use change induced by the program, which we find can vary substantially across offers. With its rental rate penalty, the CRP's Environmental Benefits Index favors offers that bring more acreage at lower cost, but this method of prioritization has the indirect effect that high EBI offers are less likely to be intensively used for commodity crops. When these effects are assessed together, we find that as more land is accepted into CRP, though the inherent environmental value of land declines, the degree to which crop acreage is displaced only decreases marginally. Thus, environmental benefits that depend on replacing crop acreage with conservation cover can be obtained even if a high amount of acreage is enrolled in CRP.

Further, the program does not currently distinguish between returning and new applicants or by land-use differences across regions, despite our finding that there is considerable heterogeneity in land use after rejection by prior CRP status and across U.S. States. Whether it is desirable for the program to account more explicitly for differences in land-use impacts will depend on the specific benefits being considered, some of which may depend on retiring specific types of land use or on the persistent use of land for conservation cover. This report finds that landowners with exiting CRP land are less likely to return to intensive commodity cropping compared to those making new offers. This finding suggests that for some environmental outcomes, prioritizing new applicants could mean retiring land likely to have been used for crops— while at the same time, taking advantage of the benefits of the less intensive uses of exiting CRP land. If instead, reenrollments were prioritized, the less intensive land uses likely to occur after CRP enrollment ends would be replaced by CRP land covers. This outcome could be desirable if the goal is long-term enrollment of individual fields.

There are several ways that the General Signup could incentivize new enrollments, including decreasing maximum rental rates for prospective reenrollments or rewarding new enrollments through the EBI. The EBI and allowed rental rates may also be used to prioritize new enrollments with a history in specific crops, such as corn or soybeans. On the other hand, the General Signup could incentivize reenrollments, such as providing additional EBI points for reenrolling in existing cover practices. More broadly, the General Signup could separately rank and accept offers for new and returning applicants, with a target ratio of new land and reenrollments. This would allow for complete flexibility to determine the program's priorities over new land and reenrollments.

Questions such as how expanding the CRP might impact the landscape, how much it will cost to expand CRP, and how these impacts differ for different subpopulations might be informative when considering expanding acreage in the CRP. However, this report has important limitations in providing insights for CRP changes. First, although our results are useful for incremental expansions of CRP Signups that would include similar offers, the expansions may not carry over to large-scale increases in the CRP or changes in program design. Such changes may result in a set of offers that differs significantly from those being assessed in this report. Second, the report leaves remaining questions as to the long-run land use impacts of the CRP. Land-use decisions beyond the immediate years after the 2016 Signup are not observed in this report. Further, although indirect land-use impacts of the CRP are a subject of great interest (e.g., Wu, 2000; Fleming, 2014), this report focuses only on the direct land-use impacts of CRP.

Appendix A: Values Corresponding with Figures in the Main Text

The following appendix provides values corresponding to the Sankey diagram (figure 2) and U.S. State-level figures in the main text and appendix (figures 10, E.2, 13, and E.3). Table A.1 presents a detailed breakdown of the share of acreage planted to a specific crop or land use before and after the 2016 General Signup among fields rejected from that enrollment period. The most common crop was wheat both before and after the Signup, and the share in each crop rose after rejection. Cropping was the most common non-CRP land use before and after rejection while grass was the next most common non-CRP land use before and after rejection. Notably, all non-CRP land uses increased in share of acreage after rejection.

Table A.2 provides the proportions of acres in each post-Signup land use among the rejected offers split out by State. Table A.3 also provides the proportions of acres in each post-Signup land use among the rejected offers split out by State but excludes offers with land that is subsequently in CRP through Continuous Signup.

Table A.4 provides the potential acreages of each land use that could be retired per $100 in rental payments across all rejected offers split by State. Table A.5 provides a breakdown of the EBI scores of rejected offers by State. Table A.5 also provides totals for acreage offered and potential rents that could be paid among rejected offers by State. Finally, table A.6 provides proportions of acreage by intended use for post-Signup acres in grass and mixed forage among parcels with rejected offers.

Table A.1. Share of acreage by land use before and after the 2016 CRP General Signup, among fields affiliated with offers submitted to that Signup

Fields rejected in 2016			Fields accepted in 2016	
	2013–16	2017–19	2013–16	2017–19
Wheat	10.6	13.7	8.14	0.07
Soybeans	7.83	8.85	3.18	0.01
Corn	6.21	7.74	1.67	0.01
Sorghum	1.02	1.28	0.65	0.00
Alfalfa	0.58	0.71	0.31	0.00
Barley	0.42	0.52	0.20	<0.01
Oats	0.25	0.28	0.07	<0.01
Other crops	2.41	4.63	3.10	0.06
Mixed forage	5.73	14.7	5.09	0.03
Grass	7.60	20.7	7.83	0.25
Idle	0.02	0.88	0.01	0.08
Fallow	6.42	9.73	5.35	0.15
Timber	0.02	0.58	0.01	<0.01
CRP	50.8	15.7	64.4	99.3

Source: USDA, Economic Research Service analysis of USDA, Farm Service Agency Crop Acreage Reporting database.

CRP=USDA's Conservation Reserve Program.

Note: Acreage shares are shown by crop planted or land use within each time period. An individual field may contribute to multiple categories, proportional to each land use or crop. Rejected fields include only those rejected and not able to enroll. Accepted fields include only those accepted and enrolled. Among accepted offers, non-CRP land uses stem from those fields that were enrolled and then exited.

Table A.2. Proportion of acres in land uses for rejected offers, after CRP Signup 49, by U.S. State

State	Corn	Soy-beans	Wheat	Other crops	Timber	Fallow or idle	Mixed forage	Grass	CRP
Alabama	0.0127	0.0174	0.0046	0.1420	0.4425	0.1009	0.0000	0.1690	0.1110
Arkansas	0.0000	0.0141	0.0000	0.0005	0.2552	0.2523	0.1898	0.0236	0.2644
Colorado	0.0718	0.0000	0.1245	0.0758	0.0000	0.1270	0.1988	0.2660	0.1361
Florida	0.0124	0.0000	0.0083	0.4689	0.0274	0.0758	0.0000	0.0290	0.3781
Georgia	0.0022	0.0287	0.0045	0.0933	0.0000	0.3870	0.0000	0.0165	0.4679
Idaho	0.0000	0.0000	0.1099	0.0585	0.0070	0.0715	0.0487	0.3494	0.3550
Illinois	0.1897	0.2519	0.0107	0.0149	0.0021	0.0169	0.0293	0.0552	0.4293
Indiana	0.1478	0.1955	0.0124	0.0054	0.0566	0.0505	0.0477	0.0716	0.4124
Iowa	0.2121	0.1901	0.0002	0.0410	0.0073	0.0024	0.0430	0.0231	0.4809
Kansas	0.1037	0.0588	0.1632	0.0942	0.0000	0.1112	0.0010	0.3748	0.0932
Kentucky	0.2388	0.2599	0.0568	0.0080	0.0000	0.0294	0.0408	0.0274	0.3391
Louisiana	0.0316	0.1423	0.0000	0.0728	0.2709	0.0196	0.0000	0.0000	0.4627
Michigan	0.1874	0.2119	0.0370	0.0264	0.0007	0.0353	0.0043	0.1350	0.3621
Minnesota	0.2835	0.3009	0.0397	0.0643	0.0132	0.0125	0.0462	0.0581	0.1817

Table A.2. (Continued)

State	Corn	Soy-beans	Wheat	Other crops	Timber	Fallow or idle	Mixed forage	Grass	CRP
Mississippi	0.0244	0.0946	0.0068	0.0554	0.1923	0.0104	0.0041	0.0645	0.5476
Missouri	0.1811	0.3957	0.0174	0.0184	0.0001	0.0118	0.1159	0.0308	0.2288
Montana	0.0003	0.0055	0.2714	0.0969	0.0000	0.1582	0.3934	0.0240	0.0503
Nebraska	0.1945	0.1381	0.0597	0.0567	0.0013	0.0582	0.0021	0.3225	0.1668
New Mexico	0.0000	0.0000	0.0515	0.0261	0.0000	0.0280	0.0000	0.5662	0.3282
New York	0.1496	0.0221	0.0115	0.0180	0.0000	0.1480	0.1753	0.4681	0.0073
North Carolina	0.0681	0.0921	0.0088	0.0390	0.1467	0.3794	0.0000	0.2001	0.0657
North Dakota	0.1134	0.2738	0.1606	0.1247	0.0001	0.0389	0.1689	0.0173	0.1024
Ohio	0.1724	0.3376	0.0390	0.0126	0.0068	0.0602	0.0679	0.0715	0.2320
Oklahoma	0.0082	0.0138	0.0961	0.0377	0.0007	0.0430	0.0026	0.7806	0.0172
Oregon	0.0013	0.0000	0.2698	0.0172	0.0000	0.2837	0.1813	0.2248	0.0218
South Carolina	0.0937	0.3046	0.0371	0.1264	0.3269	0.1113	0.0000	0.0000	0.0000
South Dakota	0.1349	0.1429	0.0423	0.1071	0.0015	0.0138	0.1716	0.1503	0.2356
Tennessee	0.1463	0.1670	0.0176	0.0200	0.0453	0.0121	0.0573	0.0723	0.4621
Texas	0.0050	0.0000	0.0618	0.1465	0.0000	0.0564	0.2335	0.3785	0.1182
Utah	0.0000	0.0000	0.1325	0.0766	0.0000	0.2530	0.0402	0.4872	0.0105
Washington	0.0008	0.0000	0.2900	0.0339	0.0004	0.2451	0.1329	0.0470	0.2501
Wisconsin	0.2538	0.1820	0.0156	0.0862	0.0326	0.0223	0.0540	0.1020	0.2516
Wyoming	0.0000	0.0000	0.0021	0.0016	0.0000	0.1111	0.4921	0.3605	0.0327

Source: USDA, Economic Research Service analysis of USDA, Farm Service Agency Crop Acreage Reporting data and CRP offer data.

CRP=USDA's Conservation Reserve Program.

Note: This table corresponds with figure 10 in the report. The table shows proportions of acres in each post-Signup land use for rejected offers by U.S. State. The "CRP" category almost entirely includes Continuous Signup enrollments. Not all States are included in the analysis. Some States are omitted due to either insufficient or no offers submitted in 2016, or lack of observations after linking offers to land use data.

Table A.3. Proportion of acres in land uses for rejected offers after CRP Signup 49, by U.S. State (excluding Continuous CRP)

State	Corn	Soybeans	Wheat	Other crops	Timber	Fallow or idle	Mixed forage	Grass
Alabama	0.0143	0.0195	0.0051	0.1597	0.4977	0.1135	0.0000	0.1901
Arkansas	0.0000	0.0192	0.0000	0.0007	0.3469	0.3430	0.2581	0.0321
Colorado	0.0832	0.0000	0.1441	0.0877	0.0001	0.1470	0.2301	0.3079
Florida	0.0200	0.0000	0.0133	0.7541	0.0440	0.1219	0.0000	0.0467
Georgia	0.0041	0.0539	0.0085	0.1753	0.0000	0.7272	0.0000	0.0310
Idaho	0.0000	0.0000	0.1703	0.0908	0.0108	0.1108	0.0756	0.5417
Illinois	0.3323	0.4414	0.0188	0.0261	0.0037	0.0297	0.0513	0.0968
Indiana	0.2515	0.3328	0.0212	0.0091	0.0963	0.0859	0.0812	0.1219

State	Corn	Soybeans	Wheat	Other crops	Timber	Fallow or idle	Mixed forage	Grass
Iowa	0.4085	0.3662	0.0004	0.0790	0.0140	0.0046	0.0829	0.0444
Kansas	0.1144	0.0648	0.1800	0.1039	0.0000	0.1226	0.0011	0.4133
Kentucky	0.3614	0.3932	0.0859	0.0121	0.0000	0.0444	0.0617	0.0414
Louisiana	0.0588	0.2649	0.0000	0.1355	0.5043	0.0366	0.0000	0.0000
Michigan	0.2938	0.3322	0.0580	0.0413	0.0011	0.0553	0.0067	0.2116
Minnesota	0.3464	0.3677	0.0485	0.0786	0.0161	0.0153	0.0564	0.0710
Mississippi	0.0540	0.2091	0.0149	0.1225	0.4250	0.0230	0.0090	0.1426
Missouri	0.2348	0.5131	0.0226	0.0238	0.0001	0.0153	0.1503	0.0399
Montana	0.0003	0.0058	0.2857	0.1020	0.0000	0.1666	0.4142	0.0253
Nebraska	0.2335	0.1658	0.0717	0.0681	0.0015	0.0699	0.0025	0.3871
New Mexico	0.0000	0.0000	0.0767	0.0388	0.0000	0.0416	0.0000	0.8428
New York	0.1507	0.0223	0.0116	0.0182	0.0000	0.1491	0.1765	0.4716
North Carolina	0.0729	0.0986	0.0094	0.0418	0.1570	0.4061	0.0000	0.2142
North Dakota	0.1263	0.3050	0.1789	0.1389	0.0001	0.0433	0.1882	0.0193
Ohio	0.2245	0.4396	0.0507	0.0164	0.0089	0.0784	0.0884	0.0931
Oklahoma	0.0084	0.0141	0.0977	0.0384	0.0007	0.0438	0.0026	0.7943
Oregon	0.0014	0.0000	0.2758	0.0175	0.0000	0.2900	0.1854	0.2298
South Carolina	0.0937	0.3046	0.0371	0.1264	0.3269	0.1113	0.0000	0.0000
South Dakota	0.1765	0.1869	0.0553	0.1401	0.0020	0.0180	0.2245	0.1966
Tennessee	0.2719	0.3105	0.0326	0.0372	0.0843	0.0225	0.1066	0.1344
Texas	0.0056	0.0000	0.0701	0.1662	0.0000	0.0640	0.2648	0.4292
Utah	0.0000	0.0000	0.1339	0.0774	0.0000	0.2557	0.0406	0.4924
Washington	0.0010	0.0000	0.3867	0.0452	0.0005	0.3268	0.1772	0.0627
Wisconsin	0.3391	0.2432	0.0209	0.1151	0.0436	0.0298	0.0721	0.1363
Wyoming	0.0000	0.0000	0.0022	0.0016	0.0000	0.1148	0.5087	0.3727

CRP=USDA's Conservation Reserve Program.

Note: This table corresponds with figure E.1 in the report. The table shows proportions of acres in each post-Signup land use for rejected offers by U.S. State. Fields going into Continuous Signup CRP are excluded. Not all States are included in the analysis. Some States are omitted due to either insufficient or no offers submitted in 2016, or lack of observations after linking offers to land use data.

Table A.4. Acres potentially retired per $100 spent on rejected offers, by U.S. State (excluding Continuous CRP)

State	Corn	Soy-beans	Wheat	Other crops	Timber	Fallow or idle	Mixed forage	Grass	Total acres
Alabama	0.0249	0.0340	0.0089	0.2779	0.8575	0.1976	0.0000	0.3309	1.7316
Arkansas	0.0000	0.0256	0.0000	0.0010	0.4638	0.4585	0.2989	0.0394	1.2872
Colorado	0.2322	0.0000	0.4102	0.2474	0.0002	0.4191	0.5932	0.8404	2.7426
Florida	0.0328	0.0000	0.0219	1.2371	0.0722	0.2001	0.0000	0.0766	1.6406
Georgia	0.0063	0.0830	0.0130	0.2588	0.0000	1.1185	0.0000	0.0477	1.5275
Idaho	0.0000	0.0000	0.3280	0.1762	0.0210	0.2116	0.1419	1.0493	1.9279
Illinois	0.1847	0.2439	0.0106	0.0146	0.0021	0.0162	0.0276	0.0506	0.5502

Table A.4. (Continued)

State	Corn	Soy-beans	Wheat	Other crops	Timber	Fallow or idle	Mixed forage	Grass	Total acres
Indiana	0.1491	0.1996	0.0127	0.0055	0.0578	0.0497	0.0487	0.0710	0.5941
Iowa	0.1952	0.1676	0.0001	0.0379	0.0069	0.0023	0.0382	0.0193	0.4675
Kansas	0.1936	0.1098	0.3050	0.1751	0.0000	0.2075	0.0018	0.6932	1.6861
Kentucky	0.1905	0.2077	0.0457	0.0063	0.0000	0.0227	0.0309	0.0187	0.5225
Louisiana	0.0917	0.3991	0.0000	0.2113	0.7864	0.0105	0.0000	0.0000	1.4989
Michigan	0.2323	0.2636	0.0461	0.0348	0.0007	0.0373	0.0057	0.1562	0.7766
Minnesota	0.2099	0.2235	0.0301	0.0472	0.0101	0.0081	0.0317	0.0398	0.6004
Mississippi	0.0639	0.2530	0.0181	0.1483	0.5042	0.0274	0.0109	0.1557	1.1815
Missouri	0.1544	0.3345	0.0147	0.0156	0.0000	0.0098	0.0976	0.0247	0.6514
Montana	0.0009	0.0164	0.7884	0.2866	0.0001	0.4711	1.1422	0.0717	2.7774
Nebraska	0.2255	0.1574	0.0691	0.0638	0.0015	0.0647	0.0019	0.3153	0.8990
New Mexico	0.0000	0.0000	0.2164	0.1095	0.0000	0.1174	0.0000	2.3110	2.7543
New York	0.2060	0.0304	0.0159	0.0249	0.0000	0.2039	0.2414	0.6448	1.3673
North Carolina	0.0973	0.1317	0.0126	0.0558	0.2031	0.5424	0.0000	0.2861	1.3291
North Dakota	0.1806	0.4349	0.2537	0.1939	0.0002	0.0612	0.2531	0.0270	1.4046
Ohio	0.1673	0.3357	0.0389	0.0125	0.0063	0.0592	0.0655	0.0653	0.7507
Oklahoma	0.0205	0.0343	0.2383	0.0936	0.0017	0.1059	0.0064	1.9222	2.4229
Oregon	0.0018	0.0000	0.3674	0.0234	0.0000	0.3866	0.2467	0.3056	1.3316
South Carolina	0.2207	0.7174	0.0874	0.2976	0.7698	0.2622	0.0000	0.0000	2.3551
South Dakota	0.1793	0.1858	0.0535	0.1198	0.0021	0.0118	0.2026	0.1799	0.9347
Tennessee	0.2457	0.2586	0.0270	0.0330	0.0769	0.0202	0.0888	0.1105	0.8607
Texas	0.0159	0.0000	0.1971	0.4677	0.0000	0.1801	0.7370	1.1914	2.7892
Utah	0.0000	0.0000	0.3652	0.2111	0.0000	0.6971	0.0776	1.3426	2.6935
Washington	0.0016	0.0000	0.6068	0.0701	0.0008	0.5140	0.2730	0.0954	1.5618
Wisconsin	0.2314	0.1594	0.0143	0.0782	0.0306	0.0197	0.0491	0.0874	0.6702
Wyoming	0.0000	0.0000	0.0080	0.0059	0.0000	0.4241	1.8790	1.1272	3.4443

CRP=USDA's Conservation Reserve Program.

Note: This table corresponds with figure 12 in the report. The table shows acreage in each land use that could be retired per $100 in rental rates if offers rejected during the 2016 Signup were accepted by U.S. State. The number of acres potentially retired per $100 is calculated by dividing the total acres in each post-Signup land use by total offered annual rental payments of rejected offers for each State and multiplying this quotient by 100. Not all States are included in the analysis. Some States are omitted due to either insufficient or no offers submitted in 2016, or lack of observations after linking offers to land use data.

Table A.5. EBI components of rejected offers, by U.S. State (excluding Continuous CRP)

State	Total acres (in 1,000s)	Total potential pay- ments (in $1,000)	EBI components						Total EBI
			N1	N2	N3	N4	N5	Cost	
Alabama	3.13	181.39	35.56	42.61	29.26	20.16	8.71	96.33	232.62
Arkansas	1.34	104.10	36.11	60.31	35.97	20.34	7.00	90.84	250.57
Colorado	84.50	3081.13	49.53	23.40	34.00	0.77	20.03	114.37	242.10
Florida	0.25	15.24	11.97	50.39	1.84	19.66	9.16	91.73	184.75
Georgia	0.70	46.09	66.33	78.13	4.09	22.40	9.94	90.30	271.20
Idaho	6.98	362.00	41.86	36.91	60.03	1.88	19.98	104.71	265.37
Illinois	19.62	3569.47	54.97	64.48	58.54	1.93	11.25	33.20	224.37
Indiana	1.95	328.57	65.21	73.73	49.92	7.27	4.13	33.77	234.02
Iowa	18.71	3998.25	56.15	57.67	80.72	11.81	18.02	11.73	236.10
Kansas	117.84	6988.09	58.27	37.67	23.83	11.32	18.55	96.71	246.35
Kentucky	3.98	761.86	57.04	72.46	63.65	7.45	3.64	21.18	225.41
Louisiana	0.68	45.10	37.43	42.10	7.69	19.48	7.52	98.79	213.01
Michigan	3.75	483.41	72.23	74.87	3.55	3.29	17.88	60.89	232.71
Minnesota	21.61	3601.55	72.43	64.61	17.14	13.52	18.12	37.67	223.49
Mississippi	4.32	365.54	28.24	61.89	32.84	17.22	7.75	94.34	242.29
Missouri	54.96	8444.92	55.81	64.03	64.27	1.62	15.83	41.86	243.41
Montana	123.96	4463.32	62.68	18.76	30.05	2.26	14.41	110.28	238.44
Nebraska	47.75	5310.15	60.65	42.61	33.43	7.26	16.87	66.48	227.30
New Mexico	2.30	83.47	25.80	38.57	72.46	0.00	18.66	108.93	264.43
New York	1.08	79.65	35.48	72.96	11.87	1.68	3.36	88.34	213.69
North Carolina	0.26	19.86	19.92	77.02	31.26	10.03	6.34	87.90	232.47
North Dakota	58.11	4138.08	61.14	35.74	17.31	1.63	15.94	89.06	220.82
Ohio	2.83	377.78	72.31	70.62	8.77	6.60	8.21	63.44	229.97
Oklahoma	35.09	1448.06	50.07	42.42	18.70	0.56	18.67	110.92	241.35
Oregon	76.51	5745.09	45.57	52.32	34.64	4.51	22.59	98.54	258.17
South Carolina	0.36	15.33	45.43	41.15	27.76	15.71	7.69	106.58	244.32
South Dakota	22.66	2423.65	69.92	43.64	14.84	2.61	13.63	67.62	212.26
Tennessee	2.15	249.81	53.67	52.63	52.86	4.01	3.94	62.57	229.68
Texas	128.19	4596.25	52.61	25.57	24.98	0.11	21.87	114.35	239.49
Utah	6.68	247.93	60.33	53.04	9.73	0.00	23.99	113.58	260.67
Washing- ton	74.80	4790.93	46.07	56.39	34.73	0.08	24.66	101.57	263.50
Wisconsin	7.97	1191.32	49.39	69.34	39.20	4.03	10.44	51.72	224.12
Wyoming	6.08	176.67	11.39	14.68	58.50	0.11	18.23	119.89	222.81

Source: USDA, Economic Research Service analysis of USDA, Farm Service Agency CRP offer data.

CRP=USDA's Conservation Reserve Program. EBI=Environmental Benefits Index.

Note: This table corresponds with figure E.2 in the report. The table decomposes EBI scores for all rejected offers from the 2016 Signup by U.S. State. All EBI values are acreage-weighted averages for offers from the relevant State. The "Cost" component of EBI scores is based on the proposed rental payments for offers. This component will be lower for offers with higher rental rates, all things being equal. The rest of the EBI score is composed of the estimated environmental contributions of offers based on either the inherent qualities of the land being offered or on the practices being proposed in the offer. This remainder is made up of scores for impacts on air quality, enduring benefits, erosion, water quality, and wildlife habitat. N1 is the wildlife factor. N2 is the factor for water quality benefits from reduced erosion, runoff, and leaching. N3 is the erosion factor. N4 is the enduring benefits factor. N5 is the fac tor for air quality benefits from reduced wind erosion. N6 is the cost factor. The total EBI is the sum of N1 through N5, plus the Cost factor. In the first two columns of the table are total acres and total potential payments for rejected offers, respectively. Not all States are included in the analysis. Some States are omitted due to either insufficient or no offers submitted in 2016, or lack of observations after linking offers to land use data.

Table A.6. Intended uses of acreage in grass and mixed forage during 2017 through 2019, returning applicants rejected from the 2016 CRP General Signup

Fields with any prior CRP			Fields with CRP expiring in 2016	
	Grass	Mixed forage	Grass	Mixed forage
Grazing	50.7	50.1	48.4	54.3
Forage	9.62	26.5	11.6	23.4
Left standing	37.1	19.8	37.2	18.7
All other uses	1.19	1.51	1.24	1.29
Percent of total	27.3	19.3	26.2	17.5

Source: USDA, Economic Research Service analysis of USDA, Farm Service Agency Crop Acreage Reporting data and CRP offer data.

CRP=USDA's Conservation Reserve Program.

Note: Acreage shares are shown by intended use within the broader crop planted. An individual field may contribute to multiple categories, proportional to each use. The share of acreage in grass or mixed forage, (in total) is shown in the final row for context. Fields with any prior CRP are those fields offered to and rejected from the 2016 General Signup that had CRP prior to the 2017 crop year.

Fields with CRP expiring in 2016 were offered to and rejected from the 2016 General Signup, and the fields' CRP contract expired at the end of the 2016 crop year. Column totals may differ from 100 percent due to rounding.

Appendix B: Comparisons Across General Signups

Figure B.1 illustrates the full distribution of EBI scores for offers submitted to each General Signup from 2010 through 2020. As shown in panel A, the distribution of EBI scores in 2016 was lower than in all other General Signups from 2010 through 2020. Notably, the EBI formula was changed for 2020 to increase points for the same rental rate, and rental rates were changing during this period. Panel B normalizes EBI scores to account for national rental rate changes and the ways in which the EBI did or did not account for these changes. In this context, the 2016 General Signup offer distribution is particularly similar to the 2013 distribution.

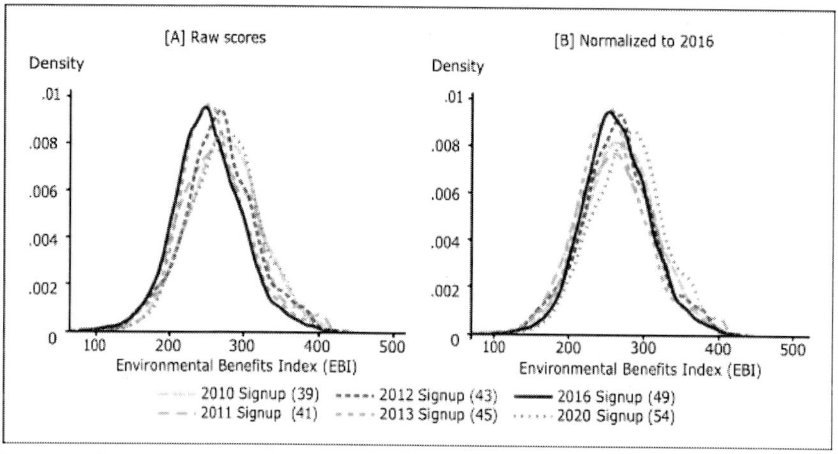

Source: USDA, Economic Research Service using data from USDA, Farm Service Agency, Conservation Reserve Program offer and contracts data, 2010, 2011, 2012, 2013, 2016, and 2020 General Signups.

Note: Kernel density plots of offers by Environmental Benefits Index (EBI) score for General Signups in 2010, 2011, 2012, 2013, 2016, and 2020. Density estimation uses an Epanechnikov kernel. Panel A plots the density for the actual EBI scores of offers in each Signup. Panel B plots the density for EBI scores normalized to account for national time trends in rental rates and the ways in which the EBI did or did not adjust for these trends at the time.

Figure B.1. Distribution of observed offer scores, 2010–20.

Figure B.2 makes comparisons of several field-level characteristics for offers made in Signups 45 and 49. The figure shows the average values for six different field-level characteristics for each EBI level across the range of EBI values for rejected offers in 2016. These values include maximum rental rates, soil pH, soil organic carbon, the National Commodity Crop Productivity Index, exogenous EBI, and erodibility. Maximum rental rates and exogenous EBI scores come from offer information, where the exogenous EBI is the total EBI excluding the points rewarded for rental rate discounts and practice choices. The rest of the variables come from USDA, Natural Resources Conservation Service Soil Survey Geographic database.

It is clear from figure B.2 that (for the most part) these Signups have similar levels of each variable across a range of EBI levels despite the differences discussed between these Signups in table 1. When combined with the information from figure B.1, which shows the EBI distributions to be similar, the information in B.2 implies that the sets of offers between the Signups are similar. This finding gives further confidence that land- use impacts will be similar for offers at similar points in the EBI distribution for different Signups.

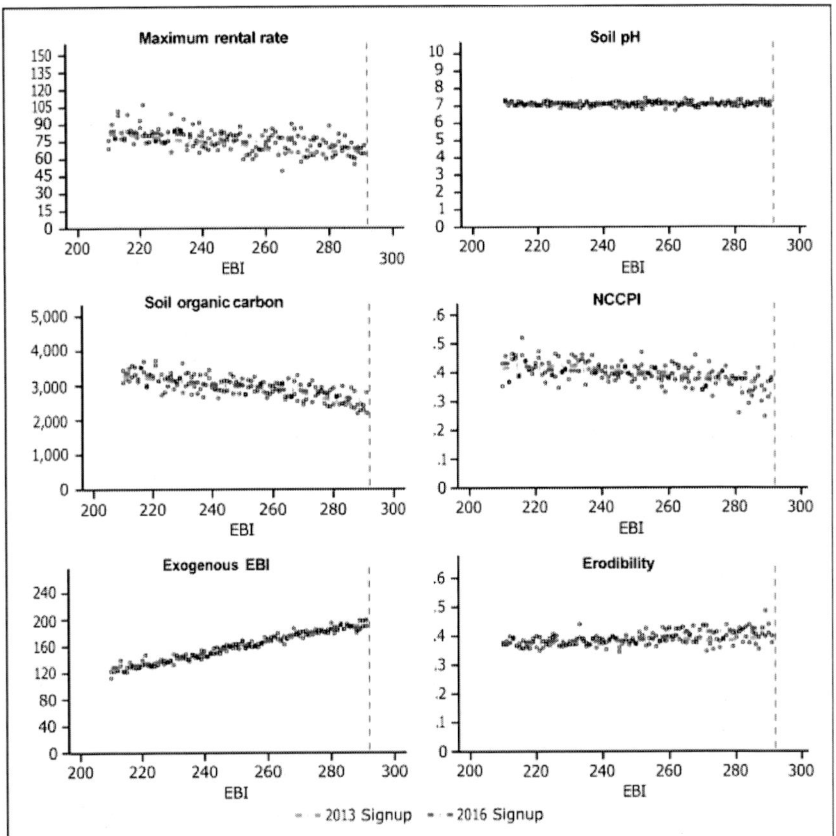

Source: USDA, Economic Research Service analysis of USDA, Farm Service Agency CRP offer and contracts data, 2013 and 2016 General Signups and USDA, Natural Resources Conservation Service SSURGO database.

NCCPI=National Commodity Crop Productivity Index. SSURGO=Natural Resources Conservation Service Soil Survey Geographic database. EBI=Environmental Benefits Index. pH=potential of hydrogen. CRP=USDA's Conservation Reserve Program.

Note: This figure compares field-level characteristics for offers made in Signups 45 and 49 in 2013 and 2016, respectively. This comparison includes maximum rental rates and exogenous EBI scores associated with offers from USDA, FSA offer data. Exogenous EBI scores involve all EBI points, except those for rental rate discounts and practice choice. The figure also includes comparisons of soil pH, soil organic carbon (0 to 20 centimeters), the National Commodity Crop Productivity Index, and the erodibility factor from the NRCS SSURGO database. For each characteristic, values were averaged for each EBI level and Signup and graphed, including local linear trends of these averages. Only offers with EBI scores below 292 (the 2016 EBI cutoff) and above 209 (the 2013 EBI cutoff) were included.

Figure B.2. Average values of field characteristics by EBI for the 2013 and 2016 CRP General Signups.

In general, as we saw in table 1, there were differences across Signups in overall acreage enrolled. Although much of these differences likely resulted from changes in the acreage cap and gaps between Signups, the differences may also result from market factors. For example, Lark et al. (2022) attributed shifts in CRP enrollment to the renewable fuel standards. Figure B.3 shows the acres offered in the 2016 Signup (panel C), as well as changes in acres offered across space from Signups 45 (in 2013) to 49 (in 2016) and from Signups 49 to 54 (in 2020). Panel C shows the highest amount of offers in Colorado, Kansas, Montana, Texas, and Washington. From 2013 to 2016, there were large increases in offered acreage in Oregon, Nebraska, Missouri, Kansas, and Washington, and large decreases in Idaho, Illinois, Texas, and New Mexico. From 2016 to 2020, there were increases in offered acreage in most States. Only Montana had a large decrease in offered acreage.

Appendix C: Comparing Land-Use Choices of Voluntary and Involuntary Exits From the CRP

Figure C.1 replicates figure 14 from the main text but excludes those fields that ultimately did enroll in CRP through the Continuous Signup. Removing the fields that enrolled in CRP provides a more direct analysis of voluntary and involuntary exits, as all of this acreage is really exiting the program. Note that voluntary exits (fields with expiring CRP acreage in 2016 that made no offer in 2016 and did not enroll in CRP by 2019) are statistically more likely to plant commodity crops than involuntary exits, which are fields with expiring CRP acreage in 2016 that made an offer in 2016 but were not accepted and did not enroll in CRP by 2019. Despite this, the difference in the proportion cropped does not fully support a hypothesis that voluntary exits primarily leave to plant crops, as over half of the acreage voluntarily leaving the program at this juncture was in grass or forage between 2017 and 2019.

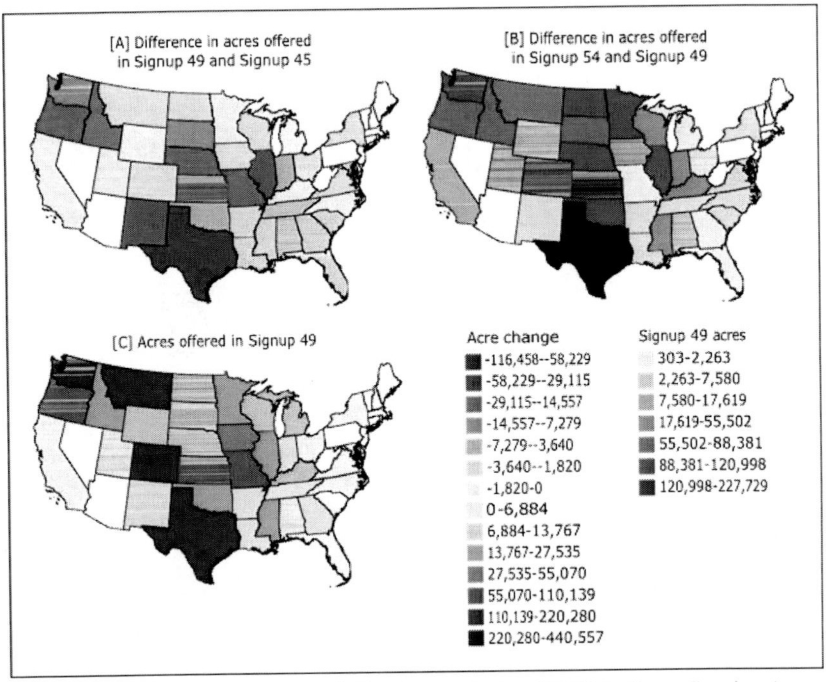

Source: USDA, Economic Research Service analysis of USDA, Farm Service Agency CRP offer data, General Signups 45, 49, and 54 (2013, 2016, and 2020).

CRP=USDA's Conservation Reserve Program.

Note: Panels A and B illustrate the difference in acreage offered to CRP between successive General Signups. Panel A presents the difference between acreage offered to Signup 49 in 2016 and Signup 45 in 2013 by State. Panel B presents the difference between acreage offered to Signup 54 in 2020 and Signup 49 in 2016 by State. Increases over time are shown in blue and decreases in red. Panel C presents the spatial distribution of offers in Signup 49. Alaska and Hawaii are visually omitted from all panels, as there were no fields offered in Signup 49 from these States. However, there were 10,148 acres offered in Alaska in Signup 54.

States in white are omitted due to either insufficient or no offers submitted in 2012, 2016, or 2020.

Figure B.3. Distribution of offered acreage in Signup 49 and differences with CRP Signups 45 and 54.

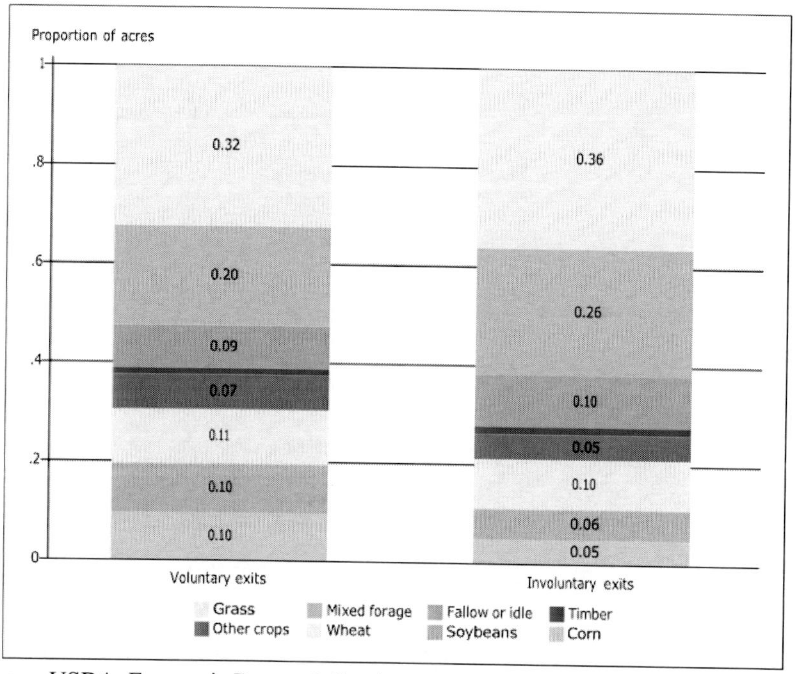

Source: USDA, Economic Research Service analysis of USDA, Farm Service Agency Crop Acreage Reporting data and CRP offer and contract data.

CRP=USDA's Conservation Resource Program.

Note: Post-Signup (2017–19) land use is weighted by acreage of fields, including only those fields with an expiring CRP contract in 2016 and no CRP enrollment in 2017 through 2019. Specific values for the proportion of acres in each post-Signup land use are reported in the figure. However, values under 0.03 are not shown due to lack of space. Proportions across categories add to 1, but re- ported values may not due to omitted values or rounding. Voluntary exits are defined within that population as fields without an offer to the 2016 General Signup. By contrast, involuntary exits are defined, within the expiring contract population, as fields with an offer to the 2016 General Signup that was rejected and no subsequent CRP enrollment in 2017 through 2019. An individual field may contribute to more than one land-use category, with the appropriate fraction apportioned to each category. Expiring CRP is restricted to conservation practices (CPs) found in General Signup contracts: introduced grasses (CP1), native grasses (CP2), existing grasses (CP10), tree planting (CP3), hardwood tree planting

(CP3A), existing trees (CP11), legacy hardwood trees (CP32), wildlife habitat (CP4D), wildlife food plots (CP12), rare and declining habitat (CP25), and pollinator habitat (CP42).

Figure C.1. Involuntary exits from CRP and nonoffers (voluntary exits) with General CRP practice expiring in 2016.

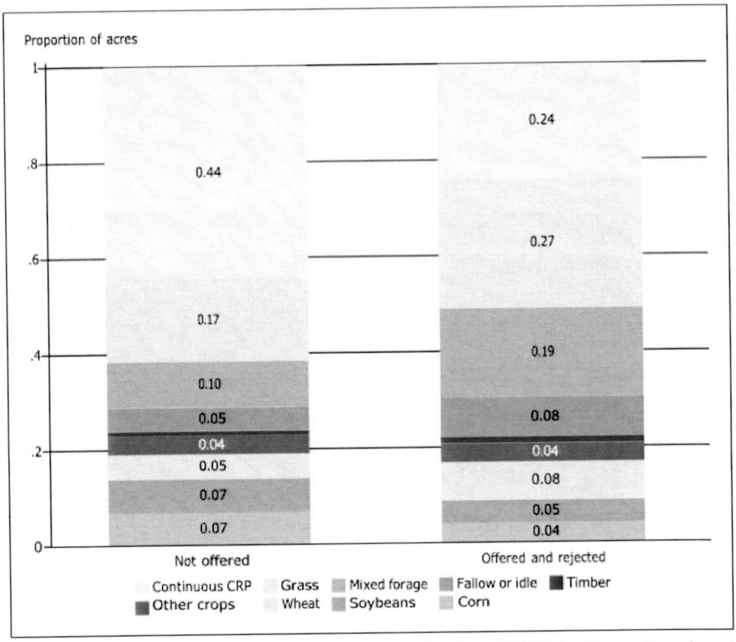

Source: USDA, Economic Research Service analysis of USDA, Farm Service Agency Crop Acreage Reporting data and CRP offer and contract data.

CRP=USDA's Conservation Reserve Program.

Note: Post-Signup (2017–19) land use is weighted by acreage of fields, including only those fields with an expiring CRP contract in 2016. Specific values for the proportion of acres in each post-Signup land use are reported in the figure. However, values under 0.03 are not shown due to lack of space. Proportions across categories add to 1, but reported values may not due to omitted values or rounding. An individual field may contribute to more than one land-use category, with the appropriate fraction apportioned to each category. Fields "Not offered" made no offer to the 2016 General Signup. Fields that are "Offered and rejected" made an unsuccessful offer in the 2016 General Signup. Acreage in Continuous Signup CRP may have enrolled with an initial contract year of 2017, 2018, or 2019. For context, the 2016 General Signup would involve an initial contract year of 2017.

Figure C.2. Post-2016 land use of fields with CRP expiring in 2016, by offer status and excluding 2016 General Signup enrollments.

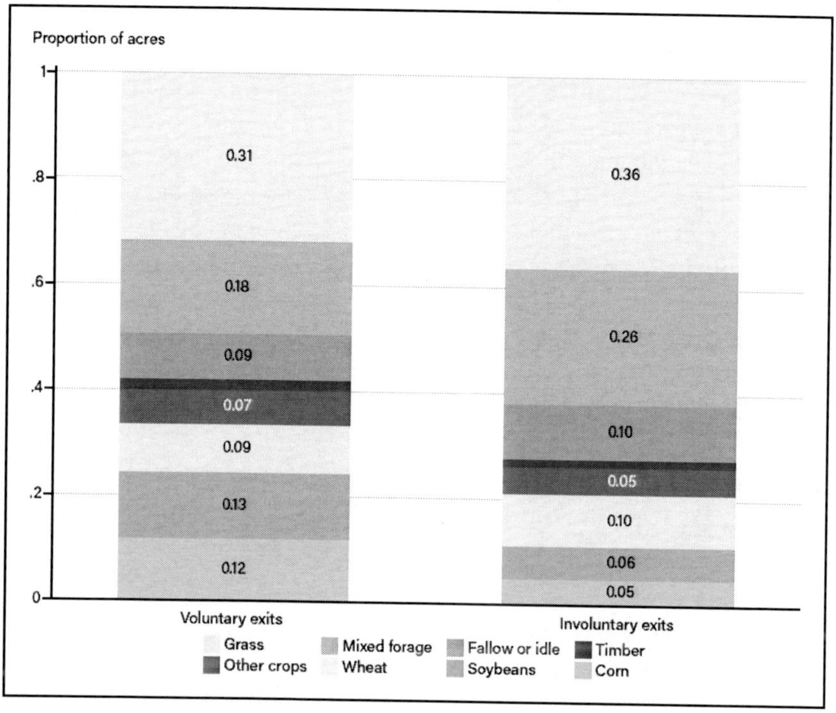

Source: USDA, Economic Research Service analysis of USDA, Farm Service Agency Crop Acreage Reporting data and CRP offer and contract data.

CRP=USDA's Conservation Reserve Program.

Note: Post-Signup (2017–19) land use is weighted by acreage of fields, including only those fields with an expiring CRP contract in 2016 and no CRP enrollment in 2017 through 2019. Specific values for the proportion of acres in each post-Signup land use are reported in the figure. However, values under 0.03 are not shown due to lack of space. Proportions across categories add to 1, but reported values may not due to omitted values or rounding. Voluntary exits are defined, within that population, as fields without an offer to the 2016 General Signup. By contrast, involuntary exits are defined, within the expiring contract population, as fields with an offer to the 2016 General Signup that was rejected and had no subsequent CRP enrollment in 2017 through 2019. An individual field may contribute to more than one land-use category, with the appropriate fraction apportioned to each category.

Figure C.3. Involuntary exits from CRP and voluntary exits with CRP expiring in 2016.

Figures C.2 and C.3 provide corollaries to figures 9 (in the report) and C.1 but include all CRP land expiring in 2016, not only that land enrolled via General Signup. As figure C.2 shows, including expiring Continuous Signup contracts increased the post-Signup proportion in Continuous Signup. This finding reflects the fact that many of these expiring Continuous Signup contracts simply elect to reenroll in the Continuous Signup. Figure C.3 provides a clearer depiction of the land-use choices of voluntary and involuntary exits from the CRP, including exits from both Continuous and General Signup contracts. The figure shows that voluntary exits were much more likely to go into crops after exiting CRP, with a slightly stronger difference than in figure C.1, potentially driven by differences in the ability to enter the Continuous Signup.

Appendix D: Comparing Rejected Offers and Nonoffered Fields

Expanding on the discussion of voluntary and involuntary exits, the authors also compared rejected offers to fields that were not offered to the 2016 General Signup. In addition to connecting the crop plantings data for offered fields, we included crop plantings data for nonoffered fields in the continental United States. Figure D.1 illustrates these land-use transitions, presenting rejected offers in panel A and fields not offered in panel B. Notably, the vast majority of fields not offered in the 2016 General Signup were in the same land use in both periods.

There were two primary ways in which the land-use transitions of fields associated with rejected offers differed substantially from fields not offered. First, as described elsewhere, these rejected offers were largely enrolled in the CRP before transitioning to non-CRP land uses after the 2016 General Signup. By contrast, those fields not offered in 2016 and engaged in the CRP prior to the Signup primarily persisted in the CRP. Second, among fields with crops between 2013 and 2016, even those fields rejected from the 2016 General Signup were more likely to enter CRP through Continuous Signup than those fields not offering to the General Signup. Presumably, those landowners offering to the General Signup were fundamentally more interested in CRP enrollment, regardless of land use.

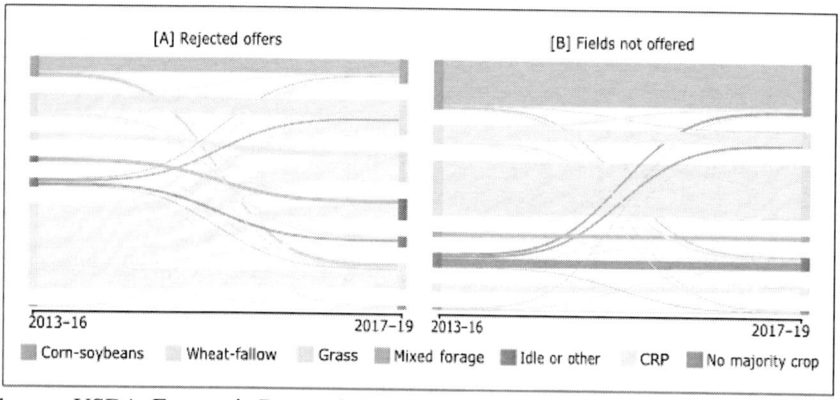

Source: USDA, Economic Research Service analysis of USDA, Farm Service Agency Crop Acreage Reporting data and Conservation Reserve Program (CRP) offer data.

Note: Each panel contains one of two subpopulations of all fields. Panel A includes those fields with an offer rejected in 2016 and no subsequent General Signup contract. Panel B includes those fields that did not offer to CRP in 2016. The total set of fields is more than 11 million.

Figure D.1. Land-use transitions among all fields, rejected offers and fields not offered in 2016.

In order to more specifically examine the differences in land-use trajectory, we conducted a matching exercise using each field's land use history, productivity potential, and county location. Specifically, we matched each field associated with a rejected offer to nonoffering fields using caliper matching. The match was based on the fraction of acre-years from 2013 through 2016 in each intended use category for a given field and the average national commodity crop productivity index (NCCPI) for a field. In order to be considered a match for a given offered field, a nonoffered field must have had a share in each category and NCCPI that was within 10 percentage points of the values for the offered field. Furthermore, fields could only be matched if they were in the same county as the field associated with the rejected offer and if each field had the same set of intended uses at any time during 2013 through 2016. Among eligible nonoffered fields, the match was chosen at random. In practice, this means that fields with a history of CRP and grain production would only be matched to fields in the same county that had a history of CRP and grain production with no forage, grazing, or other uses. The result of this stringent process was a very balanced sample, but the process also excluded 76 percent of fields with rejected offers.[98] A large majority (74 percent) of unmatched fields were those with a history of primarily CRP land cover.

Table D.1. Comparison of post-2016 land use of fields between rejected offers and nonoffered fields

Panel A	(1) Corn	(2) Soybeans	(3) Wheat	(4) Other crops
Offered	-0.00735 (0.00507)	0.00414 (0.00420)	-0.00642 (0.00461)	-0.00478 (0.00509)
Pre 2016 land use controls	Y	Y	Y	Y
Match-group fixed effects	Y	Y	Y	Y
Nonoffered mean	0.2712	0.2631	0.1617	0.1175
R-squared	0.789	0.832	0.824	0.668
Observations	18,217	18,217	18,217	18,217
Panel B	(1) Timber	(2) Idle or fallow	(3) Mixed forage	(4) Grass
Offered	0.000179*** (0.0000638)	0.00343 (0.00357)	0.0113*** (0.00245)	-0.00148 (0.00183)
Pre 2016 land use controls	Y	Y	Y	Y
Match-group fixed effects	Y	Y	Y	Y
Nonoffered mean	<0.0001	0.0700	0.0462	0.0700
R-squared	0.451	0.787	0.887	0.940
Observations	18,217	18,217	18,217	18,217

Source: USDA, Economic Research Service analysis of USDA, Farm Service Agency Conservation Reserve Program (CRP) offer data and Crop Acreage Reporting database.

Y=Yes.

* $p < 0.10$, ** $p < 0.05$, *** $p < 0.01$

Note: Each column represents a separate regression of the relevant post-2016 outcome on whether a field was offered, controlling for pre-2016 land use and match-group fixed effects. Standard errors in parentheses are clustered by match group but estimated independently within each regression.

Table D.1 presents eight separate regressions of land-use outcomes in 2017 through 2019 on an indicator of whether a field was offered to the 2016 General Signup, controlling for land use in 2013 through 2016. This sample includes only fields with rejected offers and nonoffered fields matched to those fields with rejected offers. All regressions included fixed effects for the match group, and standard errors were clustered by match group. The coefficients represented the statistical difference in subsequent land use

[98] Specifically, there are 35,792 fields linked to a rejected offer. Of these, 8,672 are matched to at least 1 nonoffering field. This sample excludes 27,120 fields. Excluding rejected offers that ultimately receive a General Signup contract, there are 35,594 fields linked to a rejected offer without a contract. Of these, 8,664 are matched to at least 1 nonoffering field. This sample excludes 26,930 fields.

between otherwise comparable fields, depending on whether the field was offered to the 2016 General Signup. CRP land use was the omitted category.

Table D.1 shows that offered fields were statistically more likely to engage in timber and mixed forage systems. Despite these statistically significant differences, the magnitude of the differences were small relative to the baseline. The largest difference was for mixed forage, with an effect of 1.13 percentage points on a base- line of 4.62 percent. All other effects were smaller than one percentage point, ranging from 0.018 percentage points to 0.735 percentage points.

Appendix E: Spatial Comparison of Land-Use Impacts

Figure E.1 provides a clearer picture of how landowners with rejected offers that did not subsequently enroll into CRP used their land. Even though States like Illinois, Iowa, and Missouri had high proportions of land going into Continuous Signup, the proportion of land in corn and soybeans in these States was very high for those landowners that did not enroll into CRP. Kansas and North Dakota had large proportions of land in wheat production after being rejected. Several States in the South had a large proportion in timber—including Louisiana, South Carolina, Mississippi, North Carolina, Alabama, and Arkansas—even though timber represents a low proportion of overall acreage for rejected offers. Values of proportions in the figure are provided in table A.3.

Figure E.2 shows how EBI scores and their components varied across States. In general, States that had higher rental rates had lower cost components of their EBI scores, shown in red in the figure. The figure also shows the average scores of the other EBI components (each of which corresponds with the main goals of the CRP) most prominently, including water quality, erosion, and wildlife. For States with offers that had lower cost components, the other components of EBI scores tended to be higher. This result was likely due to multiple factors. First, in order to justify an offer, landowners in these States need to have high scoring offers. These offers either have high-inherent EBI scores or more aggressive offers with high EBI practices. Second, it may be that offers from these States tend to have higher Environmental EBI scores, even without considering which landowners make offers. Values of proportions in the figure are provided in table A.5.

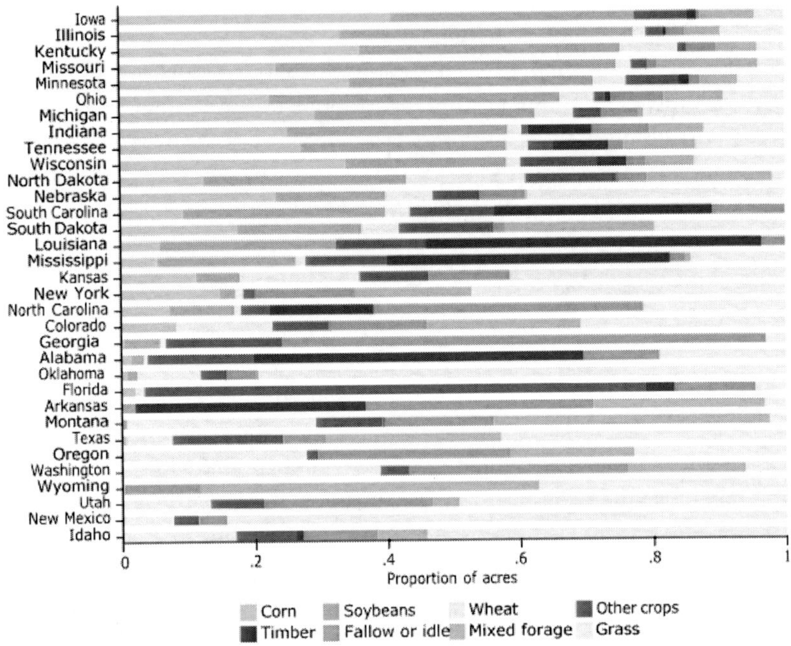

Source: USDA, Economic Research Service analysis of USDA, Farm Service Agency Crop Acreage Reporting data and CRP offer data.

CRP=USDA's Conservation Reserve Program.

Note: This figure shows proportions of acres in each post-Signup (2017–19) land use (delineated by color) for rejected offers by U.S. State. Bar segment widths (in the horizontal dimension) reflect the proportion of offered acreage in each land use within a State.

Fields going into Continuous Signup CRP are excluded. Not all States are included in the analysis. Some States are omitted due to either insufficient or no offers submitted in 2016, or lack of observations after linking offers to land use data.

Figure E.1. Proportion of acres in land uses for rejected offers, after Signup 49, by U.S. State (excluding Continuous CRP).

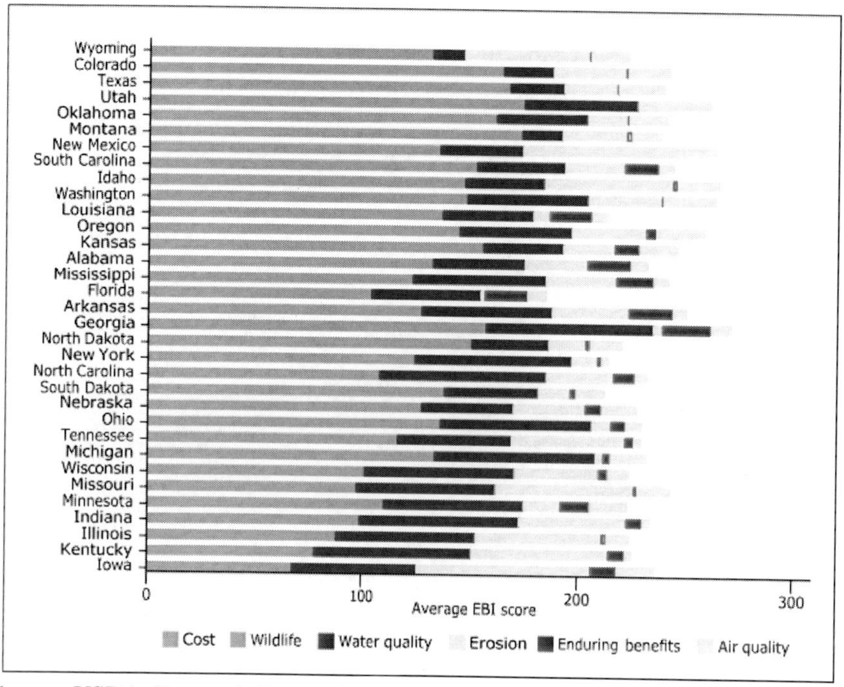

Source: USDA, Economic Research Service analysis of USDA, Farm Service Agency CRP offer data.

CRP=USDA's Conservation Reserve Program. EBI=Environmental Benefits Index.

Note: This figure decomposes EBI scores for all rejected offers from the 2016 Signup by U.S. State. The "Cost" component of EBI scores is based on the proposed rental payments for offers. This component will be lower for offers with higher rental rates, all things being equal. The rest of the EBI score comprises the estimated environmental contributions of offers based on either the inherent qualities of the land being offered or on the practices being proposed in the offer. This score is made up of scores for impacts on air quality, enduring benefits, erosion, water quality, and wildlife habitat. States are ordered by the cost component of the EBI, with States having the highest cost scores ranked first. Not all States are included in the analysis. Some States are omitted due to either insufficient or no offers submitted in 2016, or lack of observations after linking offers to land use data.

Figure E.2. EBI components of rejected offers, by U.S. State (excluding Continuous CRP).

Finally, figure E.3 provides the proportions of rejected acres that were in crop acreage and grassland after rejection, for returning and new applicants separately. The maps exclude offers that were enrolled in CRP through the

Continuous Signup. The figure reflects results from figure 7 in the report, which showed that new applicants tended to have more crop acreage than returning applicants. Figure E.3 also reflects figure 7 in that much of the grassland occurring after rejection was from returning applicants. This figure also implies that the profitability of crop conversion varied considerably. Specifically, among returning applicants, land- owners in the Midwest were much more likely to go into crops after rejection, implying that the persistence of CRP cover varies across space.

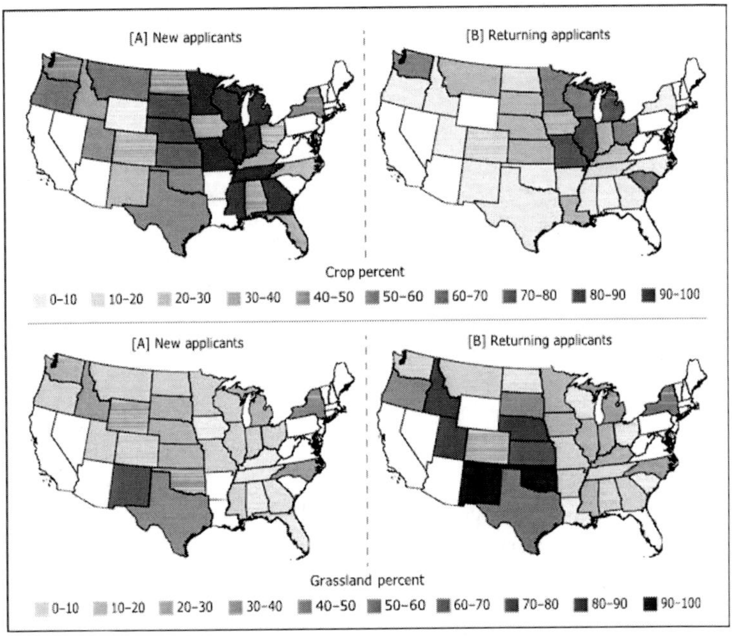

Source: USDA, Economic Research Service analysis of USDA, Farm Service Agency Crop Acreage Reporting data and USDA's Conservation Reserve Program (CRP) offer data.

Note: Each panel illustrates the percentage of crop acreage (top row) or grassland (bottom row) acres offered to and rejected from the 2016 General Signup by State. Column A shows the percentage in crop acreage (top) or grassland (bottom) for new applicants, and column B shows the percentage in crop acreage (top) or grassland (bottom) for returning applicants. Alaska and Hawaii are visually omitted from all panels, as there were no fields offered in Signup 49 from these States. States in white are omitted due to either insufficient or no offers submitted in 2016, or a lack of observations after linking offers to land use data.

Figure E.3. Crop and grassland percentages during 2017 through 2019 among fields rejected from the 2016 General Signup.

Figure E.4 illustrates the USDA, ERS farm resource regions, as referenced in the main report text for figure 13. Each U.S. county is assigned to a single resource region, with regions defined by agricultural production characteristics. For more information regarding the USDA, ERS farm resource regions, see Heimlich (2000).

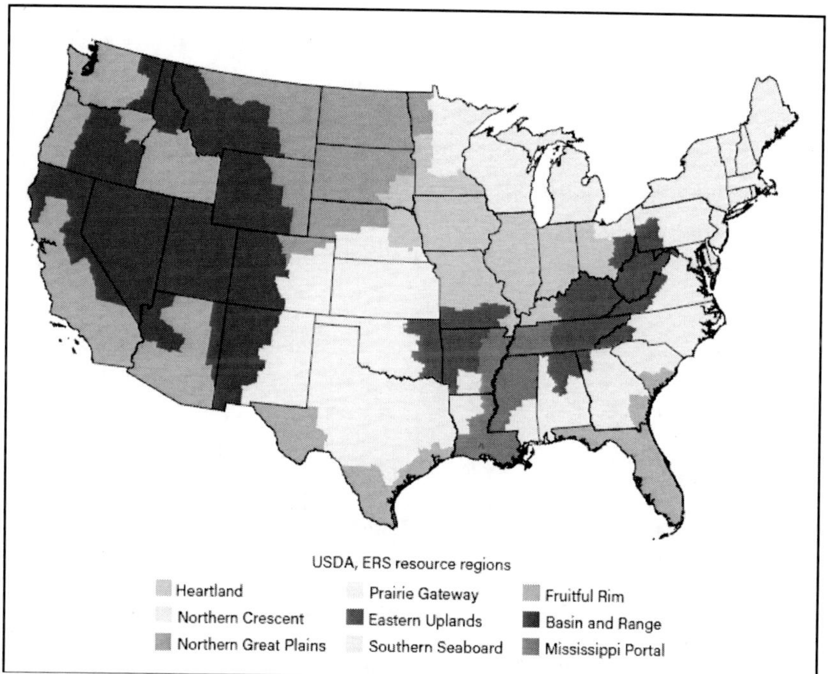

Source: USDA, Economic Research Service using Heimlich, R.E. (2000). Farm resource regions. AIB-760, U.S. Department of Agriculture, Economic Research Service.

Note: The map illustrates the county assignments of the USDA, Economic Research Service farm resource regions. Alaska and Hawaii are visually omitted, as there were no fields offered in Signup 49 from these States.

Figure E.4. USDA, Economic Research Service resource regions.

Appendix F: Proportion of Acres in Land Uses by Practice Category and Prior CRP Status

Figures F.1 through F.3 provide added detail to figure 14 from the main text. These appendix figures split out the proportions of acreage in each land-use outcome by offered practice for each of three prior CRP statuses. Figure F.1 includes offered acreage that was not in CRP at all from 2013 through 2016, that is, new applicants. Figure F.2 includes offered acreage that had CRP land that expired from 2013 through 2015. Finally, figure F.3 includes offered acreage that had CRP land that expired in 2016. The first outcome was that there were notable differences in the proportions of crop acreage between the group with no prior CRP land and the two groups with expiring CRP land. Second, it is also worth noting that 2016 expiring land was the only group among the three with significant land offering tree practices. Notably, within this last group, most of the rejected land offering a tree practice subsequently went into timber usage (35 percent) or went back into CRP through the Continuous Signup (47 percent).

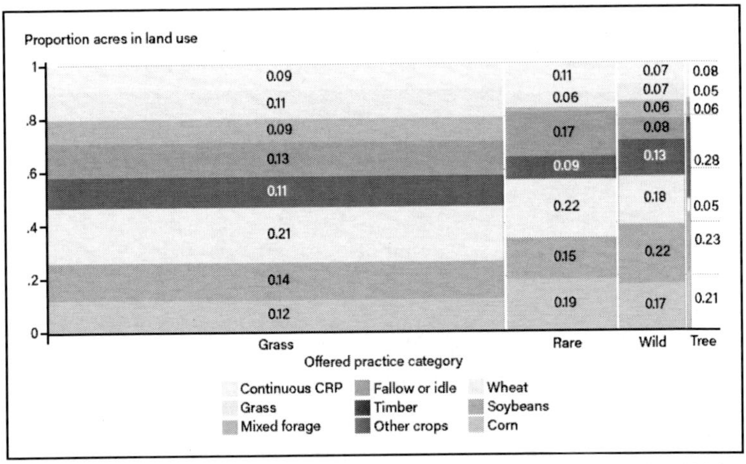

Note: This figure shows proportions of acres in each post-Signup (2017–19) land use (delineated by color) for rejected offers by offered practice category (horizontal axis). Specific values for the proportion of acres in each post-Signup land use are reported in the figure. However, values under 0.03 are not shown due to lack of space. Proportions across categories add to 1, but reported values may not due to omitted values or rounding. This figure only includes offers that had no land in CRP from 2013 through 2016. Bar widths for each practice category (in the horizontal dimension) reflect the proportion of total offered acreage in each offered category. For example, the majority of rejected acreage had offered to implement a grass practice. Bar segment heights (in the vertical dimension) reflect the proportion of acres, within offered practice category, in a given land use or crop. The grass practice category includes "Introduced grasses" and "Native grasses." The rare practice category includes "Rare and declining habitat" and "Pollinator habitat." The wild practice category includes "Wildlife habitat." The tree practice category includes "Tree planting" and "Hardwood tree planting."

Source: USDA, Economic Research Service analysis of USDA, Farm Service Agency Crop Acreage Reporting data and USDA's Conservation Reserve Program (CRP) offer data.

Figure F.1. Proportion of acres in land uses by practice category of 2016 General Signup rejected offers, for fields with no prior CRP land.

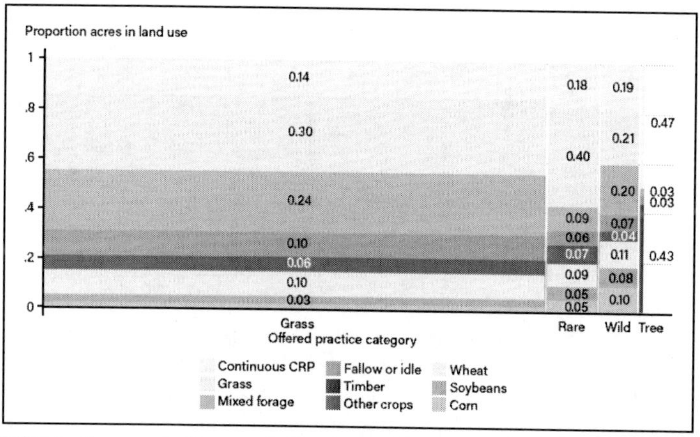

Source: USDA, Economic Research Service analysis of USDA, Farm Service Agency Crop Acreage Reporting data and USDA, Conservation Reserve Program (CRP) offer data.

Note: This figure shows proportions of acres in each post-Signup (2017–19) land use (delineated by color) for rejected offers by offered practice category (horizontal axis). Specific values for the proportion of acres in each post-Signup land use are reported in the figure. However, values under 0.03 are not shown due to lack of space. Proportions across categories add to 1, but reported values may not due to omitted values or rounding. This figure only includes offers that had land in CRP expiring between 2013 and 2015. Bar widths for each practice category (in the horizontal dimension) reflect the proportion of total offered acreage in each offered category. For example, the majority of rejected acreage had offered to implement a grass practice. Bar segment heights (in the vertical dimension) reflect the proportion of acres, within offered practice category, in a given land use or crop. The grass practice category includes "Introduced grasses" and "Native grasses." The rare practice category includes "Rare and declining habitat" and

"Pollinator habitat." The wild practice category includes "Wildlife habitat." The tree practice category includes "Tree planting" and "Hardwood tree planting."

Figure F.2. Proportion of acres in land uses by practice category of 2016 General Signup rejected offers, for fields with CRP land expiring before 2016.

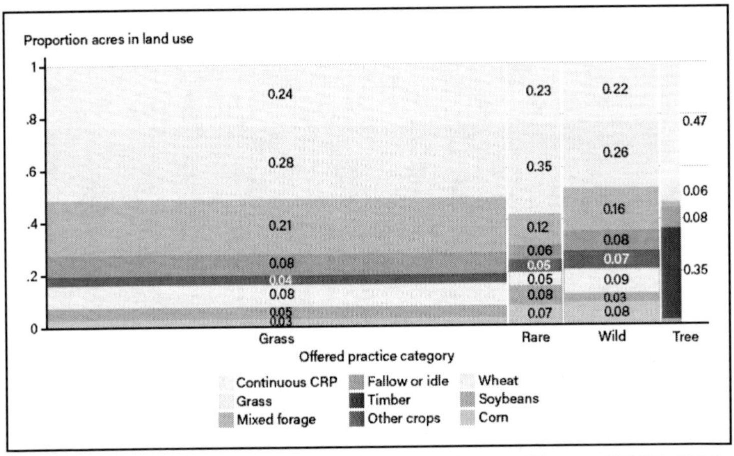

Note: This figure shows proportions of acres in each post-Signup (2017–19) land use (delineated by color) for rejected offers by offered practice category (horizontal axis). Specific values for the proportion of acres in each post-Signup land use are reported in the figure. However, values under 0.03 are not shown due to lack of space. Proportions across categories add to 1, but reported values may not due to omitted values or rounding. This figure only includes offers that had land in CRP expiring in 2016. Bar widths for each practice category (in the horizontal dimension) reflect the proportion of total offered acreage in each offered category. For example, the majority of rejected acreage had offered to implement a grass practice. Bar segment heights (in the vertical dimension) reflect the proportion of acres, within offered practice category, in a given land use or crop. The grass practice category includes "Introduced grasses" and "Native grasses." The rare practice category includes "Rare and declining habitat" and "Pollinator habitat." The wild practice category includes "Wildlife habitat." The tree practice category includes "Tree planting" and "Hardwood tree planting."

Source: USDA Economic Research Service analysis of USDA, Farm Service Agency Crop Acreage Reporting data and USDA's Conservation Reserve Program (CRP) offer data.

Figure F.3. Proportion of acres in land uses by practice category of 2016 General Signup rejected offers, for fields with CRP land expiring in 2016.

References

Barnes, J.C., Sketch, M., Gramza, A.R., Sorice, M.G., Iovanna, R., & Dayer, A.A. (2020). Land use decisions after the Conservation Reserve Program: Re-enrollment, reversion, and persistence in the southern Great Plains. *Conservation Science and Practice*, *2*(9), e254.

Bigelow, D., Claassen, R., Hellerstein, D., Breneman, V., Williams, R., & You, C. (2020). *The fate of land in expiring Conservation Reserve Program contracts, 2013–16* (Report No. EIB-215). U.S. Department of Agriculture, Economic Research Service.

Cramton, P., Hellerstein, D., Higgins, N., Iovanna, R., López-Vargas, K., & Wallander, S. (2021). Improving the cost-effectiveness of the Conservation Reserve Program: A laboratory study. *Journal of Environmental Economics and Management*, *108*, 102439.

Dodson, C., McElroy, R., Gale, F., Hanson, K., & Carlin, T. (1994). *Gauging economic impacts as CRP contracts expire. Agricultural outlook*, U.S. Department of Agriculture, Economic Research Service.

Fleming, D.A. (2014). Slippage effects of land-based policies: Evaluating the Conservation Reserve Program using satellite imagery. *Papers in Regional Science*, *93*, S167–S178.

Hansen, L. (2007). Conservation reserve program: Environmental benefits update. *Agricultural and Resource Economics Review*, *36*(2), 267–280.

Heimlich, R.E. (2000). *Farm resource regions* (Report No. AIB-760). U.S. Department of Agriculture, Economic Research Service.

Hellerstein, D., Higgins, N.A., & Roberts, M. (2015). *Options for improving conservation programs: Insights from auction theory and economic experiments* (Report No. ERR-181). U.S. Department of Agriculture, Economic Research Service.

Hendricks, N.P., & Er, E. (2018). Changes in cropland area in the United States and the role of CRP. *Food Policy*, *75*, 15–23.

Kirwan, B., Lubowski, R.N., & Roberts, M.J. (2005). How cost-effective are land retirement auctions? Estimating the difference between payments and willingness to accept in the Conservation Reserve Program. *American Journal of Agricultural Economics*, *87*(5), 1239–1247.

Lark, T.J., Hendricks, N.P., Smith, A., Pates, N., Spawn-Lee, S.A., Bougie, M., Booth, E.G., Kucharik, C.J. & Gibbs, H.K. (2022). Environmental outcomes of the U.S. renewable fuel standard. *Proceedings of the National Academy of Sciences*, *119*(9), e2101084119.

Li, C., Moore, J.M., Acosta-Martínez, V., Fultz, L.M., & Kakarla, M. (2022). Conversion of Conservation Reserve Program land back to cropland: Changes in soil carbon and nitrogen dynamics during the first 5 years. *Journal of Soil and Water Conservation*, *77*(4), 333–346.

Miao, R., Feng, H., Hennessy, D.A., & Du, X. (2016). Assessing cost-effectiveness of the Conservation Reserve Program (CRP) and interactions between the CRP and crop insurance. *Land Economics*, *92*(4), 593–617.

Morefield, P.E., LeDuc, S.D., Clark, C.M., & Iovanna, R. (2016). Grasslands, wetlands, and agriculture: The fate of land expiring from the Conservation Reserve Program in the Midwestern United States. *Environmental Research Letters*, *11*(9), 094005.

Pratt, B., & Wallander, S. (2022). *Cover practice definitions and incentives in the Conservation Reserve Program* (Report No. EIB-233). U.S. Department of Agriculture, Economic Research Service.

Ribaudo, M. (1989). *Water quality benefits from the Conservation Reserve Program* (Report No. AER-606). U.S. Department of Agriculture, Economic Research Service.

Roberts, M.J., & Lubowski, R.N. (2007). Enduring impacts of land retirement policies: Evidence from the Conservation Reserve Program. *Land Economics*, *83*(4), 516–538.

Rosenberg, A.B., & Pratt, B. (2024). Land use impacts of the Conservation Reserve Program: An analysis of rejected offers. *American Journal of Agricultural Economics*, *106*(3), 1217–1240.

Smith, R.B. (1995). The conservation reserve program as a least-cost land retirement mechanism. *American Journal of Agricultural Economics*, *77*(1), 93–105.

Sullins, D.S., Bogaerts, M., Verheijen, B.H., Naugle, D.E., Griffiths, T., & Hagen, C.A. (2021). Increasing durability of voluntary conservation through strategic implementation of the Conservation Reserve Program. *Biological Conservation*, *259*, 109177.

U.S. Department of Agriculture, Farm Service Agency. (2021). *Conservation Reserve Enhancement Program, Fact Sheet – December 2021*. U.S. Department of Agriculture, Farm Service Agency, Washington, DC.

U.S. Department of Agriculture, Farm Service Agency. (2022). *Conservation Reserve Program, Monthly Summary – September 2022*. U.S. Department of Agriculture, Farm Service Agency, Washington, DC.

Wu, J. (2000). Slippage effects of the conservation reserve program. *American Journal of Agricultural Economics*, *82*(4), 979–992.

Chapter 10

Conservation Reserve Program: Continuous Enrollment Period*

U.S. Department of Agriculture

Overview

The U.S. Department of Agriculture (USDA) is accepting offers for specific conservation practices under the Conservation Reserve Program Continuous Signup.

The USDA Farm Service Agency's (FSA) CRP is a voluntary program that contracts with agricultural producers so that environmentally sensitive agricultural land is devoted to conservation benefits. CRP participants establish long-term, resource-conserving vegetative species, such as approved grasses or trees (known as "covers"), to control soil erosion, improve the water quality and enhance wildlife habitat. In return, FSA provides participants with annual rental payments and cost-share assistance. Continuous signup enrollment contracts are 10 to 15 years in duration.

On Nov. 16, 2023, President Biden signed into law, H.R. 6363, the Further Continuing Appropriations and Other Extensions Act, 2024, which extended the 2018 Farm Bill. This extension now allows authorized programs, including Continuous CRP, to continue through Sept. 30, 2024.

---

* This is an edited, reformatted and augmented version of U.S. Dewpartment of Agriculture, Farm Service Agency Publication, dated January 2024.

In: The Conservation Reserve Program
Editor: Jerry P. Norton
ISBN: 979-8-89530-774-8

Benefits

CRP protects millions of acres of American topsoil from erosion and is designed to safeguard the nation's natural resources. By reducing water runoff and sedimentation, CRP protects groundwater and helps improve the condition of lakes, rivers, ponds and streams. The vegetative covers also make CRP a major contributor to increased wildlife populations in many parts of the country.

Administration

FSA administers CRP while technical support functions are provided by:

- USDA's Natural Resources Conservation Service (NRCS);
- State forestry agencies;
- Local soil and water conservation districts; and
- Other non-federal providers of technical assistance.

Continuous Enrollment

To submit an offer for Continuous CRP signup, producers should contact the FSA at their local USDA Service Center by July 31, 2024, in order to have an offer effective by Oct. 1, 2024. To ensure enrollment acreages do not exceed the statutory cap, FSA will accept offers from producers on a first-come, first-served basis and will return offers for approval in batches throughout the year.

Additionally, producers with acres enrolled in Continuous CRP set to expire September 30, 2024, can offer acres for re-enrollment beginning Jan. 12, 2024. A producer can both enroll new acres into Continuous CRP and re-enroll any acres expiring September 30, 2024.

To find your local FSA office visit https://offices.sc.egov.usda.gov/locator/app.

Who is Eligible?

A producer must have owned or operated the land for at least 12 months prior to submitting the offer, unless:

- The new owner acquired the land due to the previous owner's death;
- The ownership change occurred due to foreclosure where the owner exercised a timely right of redemption in accordance with state law; or
- The circumstances of the acquisition present adequate assurance to FSA that the new owner did not acquire the land for the purpose of placing it in CRP.

Eligible Land

Land must be cropland that is planted or considered planted to an agricultural commodity four of the six crop years from 2012 to 2017 and is physically and legally capable of being planted (no planting restrictions due to an easement or other legally binding instrument) in a normal manner to an agricultural commodity.

Certain marginal pastureland that may be devoted to riparian buffers, wildlife habitat buffers, or wetland buffers is also eligible.

Eligible Practices

Land must be eligible and suitable for any of the following conservation practices:

- Grass Waterway
- Shallow Water Area for Wildlife
- Contour Grass Strip
- Filter Strip
- Riparian Buffer
- Denitrifying Bioreactor on Filter Strip and Riparian Buffer
- Saturated Filter Strip and Riparian Buffer
- Habitat Buffers for Upland Birds
- Wetland & Buffer SAFE Practices
- Wetland Restoration on Floodplain and Non- floodplain
- Prairie Strips
- Windbreaks, Shelterbelts and Living Snow Fences
- Marginal Pastureland Wetland Buffer and Wildlife Habitat Buffers
- Long Leaf Pine Establishment
- Duck Nesting Habitat
- Pollinator Habitat
- Bottomland Timber Establishment on Wetlands
- Farmable Wetlands Program (FWP) Constructed Wetland
- FWP Aquaculture Wetland Restoration
- FWP Flooded Prairie Wetland
- Farmable Wetlands and Farmable Wetland Buffer
- Wellhead Protection Area Practices
- State Acres for Wildlife Enhancement (SAFE)
- Highly Erodible Land Initiative (HELI)

Payments

Under CRP continuous signup, FSA provides eligible participants with annual rental payments and cost-share assistance.

A Signup Incentive Payment of 32.5 percent of the first full year annual rental payment is paid at contract approval on all newly enrolled Continuous CRP contract acreage. A Practice Incentive Payment of up to 50 percent is paid for installing the practice.

Annual Rental Payments

- In return for establishing long-term, resourceconserving covers, FSA provides annual rental payments at 90 percent of the existing rate to eligible CRP continuous participants based on per acre cropland and marginal pasture land rental rates, plus a 10 percent "inflationary" adjustment, determined by FSA.
- The per-acre annual rental rate may not exceed the calculated rate or $300/acre. Producers may elect to receive a rental rate amount less than the maximum payment rate.
- Climate-Smart Practice Incentive: FSA provides an incentive of 3, 5, or 10 percent for CRP practices that will increase carbon sequestration, reduce GHG emissions, and otherwise are climate smart practices. The incentive amount is based on the estimated benefits of each practice.
- Water Quality Incentive: FSA provides a 20 percent rental rate incentive for CRP water quality practices that will improve water quality by reducing sediment loading, nutrient loading and harmful algae blooms.

Cost-Share Assistance

Cost-share of up to 50 percent of the re- reimbursable cost of installing the practice is provided by FSA. Cost-share assistance is not available for any management activity.

More Information

For more information about FSA and its programs, visit www.fsa. usda.gov or contact your local FSA office. To find your local FSA office, visit: farmers.gov/service-locator.

Chapter 11

Conservation Reserve Program: State Acres for Wildlife Enhancement (SAFE) Initiative Habitat Restoration | Carbon Sequestration | Water Quality Enhancement*

U.S. Department of Agriculture

Why Choose CRP State Acres for Wildlife Enhancement?

The State Acres for Wildlife Enhancement (SAFE) Initiative restores vital habitat in order to meet high-priority state wildlife conservation goals. The Conservation Reserve Program (CRP) provides farmers and landowners with several opportunities like this to achieve many conservation goals. Through SAFE, landowners establish wetlands, grasses, and trees. These practices are designed to enhance important wildlife populations by creating critical habitat and food sources. They also protect soil and water health by working as a barrier to sediment and nutrient run-off before they reach waterways.

States may submit proposals for new or modified SAFE projects during announced proposal periods.

SAFE Practices

- CP-38A Buffers

* This is an edited, reformatted and augmented version of U.S. Dewpartment of Agriculture, Farm Service Agency Publication, dated April, 2022.

In: The Conservation Reserve Program
Editor: Jerry P. Norton
ISBN: 979-8-89530-774-8

- CP-38B Wetlands
- CP-38C Trees
- CP-38D Longleaf Pine
- CP-38E Grass

Practices are available through CRP continuous signup. Please check with your local FSA county office for signup periods for SAFE practices in your area.

SAFE Initiative Benefits

Species targeted by SAFE include those designated by the U.S. Fish and Wildlife Service as threatened or endangered including the lesser prairie chicken, those that have suffered population decline – like the New England cottontail, bobwhite quail, and grassland birds. SAFE also provides habitat for other wildlife, including deer, game birds such as pheasants, waterfowl, and pollinators that have significant social and economic value.

SAFE Initiative Impacts

- Enhances populations of high priority species by working with states to identify practices and sites that will have the largest benefit
- Improves native grassland habitats for rare, threatened, endangered, and declining species that are dependent on native prairie communities
- Restores trees, primarily hardwoods, to create habitat that will benefit species such as northern bob-white quail, the Indiana bat, ducks, and other wildlife
- Promotes carbon sequestration

Financial Benefits

- 10-15 years of annual rental payments
- Payments of up to 50% Cost-Share for practice establishment
- Practices enrolled through CRP Continuous Signup will also receive
    - A 50% Practice Incentive Payment (PIP)
    - Sign-up Incentive Payment (SIP) equal to 32.5% of first full year's annual rental payment
    - A Climate Smart Annual Practice Incentive of 3,5, or 10% based on the practice cover

More Information

For more information, contact your local service center and USDA Farm Service Agency office: farmers.gov/service- locator.

Index

F

G

H

L

M

N

P

Q

R

S